Study Guide

Microbiology

Principles and Applications

Study Guide

William Matthai

Tarrant County Junior College

Microbiology

Principles and Applications

Second Edition

Jacquelyn G. Black

Marymount University

PRENTICE HALL
Englewood Cliffs, NJ 07632

Editorial/ production supervision: *Susan Fisher*
Prepress buyer: *Paula Massenaro*
Manufacturing buyer: *Lori Bulwin*
Acquisitions Editor: *David Brake*
Supplement Acquisitions Editor: *Mary Hornby*

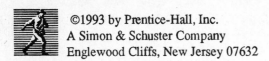

Printed in the United States of America

10 9 8 7 6 5 4 3 2 1

PRENTICE-HALL INTERNATIONAL (UK) LIMITED, LONDON
PRENTICE-HALL OF AUSTRALIA PTY. LIMITED, SYDNEY
PRENTICE-HALL CANADA INC. TORONTO
PRENTICE-HALL HISPANOAMERICANA, S.A., MEXICO
PRENTICE-HALL OF INDIA PRIVATE LIMITED, NEW DELHI
PRENTICE-HALL OF JAPAN, INC., TOKYO
SIMON & SCHUSTER ASIA PTE. LTD., SINGAPORE
EDITORA PRENTICE-HALL DO BRASIL, LTDA., RIO DE JANEIRO

TABLE OF CONTENTS

PREFACE

Microbiology is fun! It will hopefully be one of the most interesting and exciting science courses you will ever take in college because it relates to everyone and virtually everything. It involves interesting diseases and some pretty gross pictures. It involves the human body and how we defend against these diseases. It involves the hospital and clinical environments and how many of us will have to control these organisms. It even involves chemistry and genetics and how these organisms can grow and change. And it involves the environment and industry and how these organisms have become a part of our ecosystem and how we use them to our advantage.

Microbiology is always changing. I have taught this science course for over 20 years and I'm still finding something new to talk about and for that matter to learn. Some of these new discoveries, like recombinant DNA technology and monoclonal antibody production are really exciting but others like the recent AIDS epidemic and the emergence of highly resistant bacteria are rather depressing. But, they are all important parts of Microbiology. Hopefully, after you complete this course you will have a far better idea of how microbes have become an integral part of all of our daily lives. Perhaps you might even become as interested in this field as I am and maybe someday, decide to make microbiology a major part of your career plans.

THE STUDY TIPS

What you ultimately get out of a course depends a lot on how much time and effort you put in. However, everyone has differing amounts of time they can devote to any one subject. Your time is greatly limited if you have children at home, a husband or wife, work at a job outside of school time, and/or carry a full college load. Therefore you need to maximize the time you have available to you. Here are a number of "tips" you can try but please realize that what works for some, may not always work for others.

1. <u>Study in the morning.</u> Normally everyone is freshest early in the morning or when they are the least tired.

2. <u>Study in a quiet location.</u> Try the library or an empty room where distractions are at a minimum. However, if it's too quiet it may be just as bad, in which case, try a lot of constant noise like a loud fan or rock music.

3. <u>Work in a small group.</u> Sometimes it helps especially before an exam to get together with one or two friends to share notes, problems, and answers to objectives.

4. <u>Study the objectives.</u> Get a copy of the course objectives (if available) and make sure you complete all those required for the test.

5. <u>Prepare a study outline.</u> Use the outline provided in this study guide to obtain an overall idea of the material to be covered. The outlines should be brief and no more than one page per chapter.

6. <u>Study small units at a time.</u> Only those with photographic memories can incorporate whole chapters in one study session. Concentrate on <u>one</u> unit at a time, go over it several times, and then move on to the next. Memorization is usually just repetition of the facts.

7. <u>Study in small blocks of time.</u> Concentrate for 30 minutes or an hour on just a few topics then take a break. Short, concentrated "bursts" of effort are more effective than long hours of "sleepy" review.

8. <u>Look for old exams.</u> Some instructors make them available as guides. You can also use the questions provided in this study guide as a review.

9. <u>Use the text as a reference.</u> If your time is limited, scan the chapters and look at the figures. However, some instructors test "by the book" so know first how the instructor intends on generating questions.

10. <u>Know the instructor.</u> As mentioned earlier, know how they like to test and where they intend to generate their questions. Most instructors are more than willing to help and really want students to succeed. It never hurts to ask for help.

11. <u>Get a tutor.</u> Some colleges have graduate assistants or advanced students who may be able to provide assistance.

12. <u>Don't be discouraged.</u> Microbiology is fun but it's not necessarily easy. There are a lot of terms and topics that are new and, in some cases, difficult. Hang in there, take a deep breath, and keep trying.

THE LECTURE

Most microbiology courses are taught with a traditional lecture and laboratory format. With traditional lectures, note taking is essential. Here's some tips for good note taking and for getting the most out of a lecture.

1. <u>Use a separate notebook.</u> Don't mix your microbiology notes with other courses...it's too confusing.

2. <u>Start each lecture on a new page.</u> Give yourself plenty of room. Paper is not that expensive. Number and date each page.

3. <u>Abbreviate and outline your notes.</u> Don't try to write out every word. Set up major topics and indent the minor points.

4. <u>Bring a tape recorder.</u> Some instructors talk very fast!

5. <u>Use a highlighter.</u> After lecture you can highlight or underline the key points.

6. <u>Don't rewrite your notes.</u> Unless you have a computer and lots of time, this isn't really necessary.

THE TEST

Each instructor has their own method of testing. Know what kinds of questions they plan to ask. Essay tests are studied differently than multiple choice and true/false. Essays require an overall knowledge of the material and some facts, and the ability to write coherent sentences. Multiple choice and true/false questions require a knowledge of facts and details and an ability to mark with a pencil.

After the test, ask to see your results. You can always learn as much from your mistakes as from your successes.

THE TEXTBOOK

You have without a doubt the finest and best illustrated microbiology textbook available today. Study the figures, diagrams, and photos because they can provide excellent summaries of the important topics. If you don't have enough time to thoroughly read the text, scan it, or use it as a reference. The chapter summaries are excellent concentrated reviews of the chapters. Try the review questions. Your instructor has the answers. I know, because I wrote them!

THE STUDY GUIDE

This study guide was written to help any student regardless of how much time they have available. If time is limited, use only the study outlines and perhaps the learning activities and review questions. Otherwise, take advantage of all six parts. Each of these is explained below.

I. **Overview** - This is a brief summary of the entire chapter and was written to set the "tone".

II. **Study Outline** - This is very useful to help in studying for exams. Use, as is, or briefly expand each topic but keep it to only one page.

III. **Review Notes** - These notes are based on the focus questions in the front of each chapter. Figures and tables are cited and the best ones are included. Helpful tips and anecdotes are included where possible.

IV. **Learning Activities** - These include fill-in-the-blank and matching questions which come directly from the text and are usually based on the bold-faced terms.

V. **Review Questions** - These include true/false and multiple choice questions. Clinical questions are included for the disease chapters. All of the questions come directly from text material.

VI. **Answer Key** - All the answers to the learning activity and review questions are provided here. Notice that all the review answers are listed <u>with</u> an explanation <u>and</u> the page or figure reference.

As a final note, I hope that your experience with microbiology and with this study guide is a useful and enjoyable one. I have enjoyed writing it and I would like to thank Alison Munoz at Prentice Hall for her help and encouragement on the first edition and Mary Hornby for her help with the second edition, my family for putting up with me during the final production days, and my students for giving me the reasons and the incentive to always do my best.

William C. Matthai
Associate Professor
Tarrant County Junior College

CHAPTER 1

SCOPE AND HISTORY OF MICROBIOLOGY

The field of microbiology is one of the most exciting and diverse fields in all of science. It is of special interest to all those students entering the many disciplines of allied health and medicine because it provides information about microbial disease and how we can control and defend against them. It is also of value to those students entering disciplines such as food technology and environmental biology because of the important association microbes have with the foods we eat and the environment about us. In fact, microbiology, in some way, shape, or form is of interest to all of us simply because microbes are in the air we breathe, the food we eat, and the water we drink.

Microbiology is not just a study of microbes such as bacteria, fungi, and viruses. It is a study of many diverse fields including immunology, chemotherapy, and genetics. It is also a study of the people that work in all these fields, the researchers and teachers, the clinicians, and the geneticists.

Since microbiology plays such an important role in our lives, it is little wonder that this science has had a most fascinating and colorful history. The second part of the chapter presents some of the most important historical events that brought microbiology into its present form today. Beginning with the initial discovery of the bacteria by Leeuwenhoek to the development of the germ theory, the author paints a fascinating picture of the most important discoveries that provided the foundation for this science. The last part of the chapter outlines the developmental events of many of the special fields of microbiology such as immunology, virology, and chemotherapy. As can be seen, great strides have already been made in these fields and many have opened the doors to even greater discoveries.

STUDY OUTLINE

I. **Why Study Microbiology?**
   A. Microbial relationships
   B. Beneficial aspects of microbes
   C. Microbial relationships to life processes
   D. Microbes in research
II. **Scope of Microbiology**
   A. The microbes
   B. The microbiologists
III. **Historical Roots**
   A. Biblical accounts
   B. Greek and Roman contributions
   C. Bubonic plague
   D. Development of microscopy
IV. **The Germ Theory of Disease**
   A. Spontaneous generation theory
   B. Early studies
   C. Pasteur's further contributions
   D. Kochs contributions
   E. Work toward controlling infections
V. **Emergence of Special Fields of Microbiology**
   A. Immunology
   B. Virology
   C. Chemotherapy
   D. Genetics
   E. Tomorrow's history

REVIEW NOTES

A. **Why is the study of microbiology important?**

Microorganisms have always been and will continue to be a major influence in our lives. Because of their relationship to our health and welfare, it is essential that we understand their characteristics. By studying their features we can also gain insight into the processes and relationships of all life forms.

B. **What is the scope of microbiology?**

Microbiology is often considered just a study of bacteria. However, it actually is a discipline that includes not only bacteria but viruses, algae, protozoa, fungi, and even the helminths. It even includes a great variety of diverse fields such as immunology and genetics that indirectly or directly relate to these microbes. Look at Table 1.2 to obtain an overall picture of the tremendous diversity of this field of science.

C.   What are some major events in the early history of microbiology?

The field of microbiology is as old as recorded time itself, however it was never recognized as a science until late into the 19th century. The Greeks, Romans, and Jews all made major contributions to microbiology but only indirectly since microbes were not seen nor even believed to exist. These invisible organisms, however, frequently devastated major parts of the world as a result of epidemics such as the plague and smallpox and were often responsible for changing the course of history. It wasn't until the Dutch lens grinder Leeuwenhoek, using a very crude microscope, first observed the tiny "animalcules" that the relationship these organisms had to everyday activities could even be appreciated.

D.   What is the germ theory of disease and what historical developments led to its formulation?

GERM THEORY → Particular bacterium causes particular diseases.

The germ theory of disease provides us with a mechanism to relate a particular bacterium to a particular disease. Before this relationship could be established, several important concepts had to be taken care of. First and foremost, the theory of spontaneous generation, which began in the days of Aristotle, had to be disproved. This theory represented a major stumbling block in the attempts to prove that a particular bacterium could cause a particular disease. If this theory were true, bacteria and hence diseases could arise at any given time and in any given place, spontaneously. Therefore, diseases could never be prevented nor controlled. Fortunately, several scientists including Pasteur and Tyndall using very simple equipment managed to demonstrate once and for all that bacteria could only arise from other living organisms. Once this was confirmed, the stage was set for Koch to develop a series of four postulates that could be used to prove that a particular microbe could cause a particular disease.

KOCH — Proved that a particular microbe could cause a particular disease.

E.   What events mark the emergence of immunology, virology, chemotherapy, microbial genetics, and molecular biology as branches of microbiology?

With the development of the germ theory, numerous specialized fields within microbiology began to emerge. The field of immunology began with the development of the smallpox vaccine by Jenner. Virology emerged as a separate field with the discovery of pathogenic, filterable agents by Beijerinck. Ehrlich and Fleming pioneered the field of chemotherapy with the discovery of sulfa drugs and antibiotics. Lastly, the field of genetics and molecular biology was greatly expanded through the efforts of Griffith who discovered genetic transformation and Avery who demonstrated that this genetic change was due to DNA.

Jenner - developed Smallpox vaccine. (immunology)
Ehrlick & Fleming - discovered of Sulfa drugs & antibiotics (Chemotherapy)
Griffith - discovered genetic transformation (genetics)
Avery - demonstrated that this genetic change was due to DNA (genetics)

3

LEARNING ACTIVITIES

Complete each of the following items by supplying the appropriate word or phrase.

1. The study of very small organisms that require a microscope to observe them would be ___Microbiology___ .

2. Acellular entities too small to be seen even with a light microscope are ___VIRUSES___ .

3. The study of the frequency and distribution of diseases is known as ___Epidemiology___ .

4. A scientist who built a crude microscope in the 1600's and who coined the term "cell" would be ___Hooke___ .

5. Single-celled and multicellular microscopic organisms with true nuclei and which absorb nutrients from their environment are ___Fungi___ .

6. A Dutch clothes merchant and amateur lens grinder who observed the first microorganisms was ___Leeuwenhoek___ .

7. Living organisms arising from nonliving substances can be called ___Spontaneous regenaration___ .

8. An Italian physician who demonstrated that maggots arise from fly eggs and not rotten meats was ___Redi___ .

9. "Swan-necked" flasks used to refute spontaneous generation were made by ___Louis Pasteur___ .

10. The theory that states that microbes can invade other organisms and cause disease is the ___Germ Theory___ .

11. A scientist who formulated four postulates to associate a particular organism with a specific disease was ___Koch___ .

12. A Russian scientist who discovered "phagocytes" that could ingest microbes was ___Metchnikoff___ .

13. A German physician who recognized the connection between autopsies and puerperal fever was ___Semmelweiz___ .

14. The American physician credited with identifying the relationship between mosquitoes and yellow fever was ___Reed___ .

4

15. A pioneer in the development of chemotherapy and who discovered the drug Salvarsan while searching for a "magic bullet" was ____Ehrlick____.

16. A tentative explanation to account for an observed condition or event is called an ____Hypotheses____.

17. A project designed to map the location of every gene in all human chromosomes is the ____Human____ ____genome____ project.

Match the appropriate scientist with the statement listed below.

Key Choices:

a.	Schleiden & Schwann	f.	Semmelweiz
b.	Tyndall	g.	Jenner
c.	Aristotle	h.	Beijerinck
d.	Lister	i.	Pasteur
e.	Reed	j.	Domagk

__A__ 1. A Developed the cell theory.

__C__ 2. C Suggested that fire, earth, air, and water could generate life forms.

__b__ 3. b Used an airtight box to demonstrate the presence of microbes on dust particles present in the air.

__j__ 4. Developed several chemotherapeutic agents including protosil, isoniazid and other sulfanilamides.

__g__ 5. g Developed the technique of vaccination in the prevention of smallpox.

__d__ 6. b Used carbolic acid sprays and aseptic procedures in surgery.

__i__ 7. i Established that alcohol was produced in wine only in the presence of yeast.

__H__ 8. H The first scientist to characterize viruses.

Match the appropriate term with the descriptions listed below.

Key Choices:

a.	Epidemiology	f.	Mycology- Fungi
b.	Etiology	g.	Immunology
c.	Parasitology	h.	Chemotherapy
d.	Virology	i.	Genetics
e.	Algology	j.	Bacteriology

y causes é —
ns of diseases

_b_ 9. Relates to the causes of disease. _Etiology_

_d_ 10. Includes acellular structures composed of nucleic acids and a few proteins. _Virology_

_g_ 11. The study of how host organisms defend themselves against infection by microbes. _Immunology_

Virology d 12. Major contributors to this science were Chamberland and Stanley.

_F_ 13. Relates to yeasts and fungi. _Mycology_

Algology C 14. Includes those microbes with clearly defined nuclei, organelles, and mechanisms of photosynthesis.

Epidemiology A 15. A study of the frequency and distribution of diseases.

genetics i 16. Field of study based on Mendel's work.

Chemotherapy H 17. Paul Ehrlich and Alexander Fleming pioneered this field of microbiology.

REVIEW QUESTIONS

True/False (Mark T for True, F for False)

_F_ 1. Koch's postulates can be applied to the identification of all microbes including viruses.

_T_ 2. John Needham actually was a proponent of the spontaneous generation theory. _Living M.O. can arised from non-living Sub_

_F_ 3. Leeuwenhoek is credited with observing the first microorganisms and making the first microscope.

_F_ 4. Fleming is credited with the development of porcelain filters used to remove bacteria from water.

_T_ 5. The CDC stands for The Center of Disease Control.

_T_ 6. John Tyndall's contributions to microbiology was in his demonstration of microbes on dust particles.

Multiple Choice

__e__ 7. Pasteur is credited with all the following except:
    a. construction of swan-necked vessels.
    b. development of a vaccine for rabies.
    c. development of the technique of pasteurization.
    d. became the director of the Pasteur Institute in Paris, France.
    e. all of the above are credited to him.

__c__ 8. One of the most important contributions of Robert Koch in his development of the germ theory of disease was the:
    a. use of test animals in research.
    b. use of the microscope.
    c. development of the technique of pure culturing.
    d. development of the bunsen burner.

__d__ 9. Fungi can be characterized as:
    a. photosynthetic organisms.
    b. organisms lacking a cell wall.
    c. organisms lacking a true nucleus.
    d. organisms that absorb nutrients from their environment.

__d__ 10. Phycology is the study of:
    a. molds
    b. bacteria
    c. viruses
    d. algae

ANSWER KEY

Fill-in-the-Blank

1. Microbiology   2. Viruses   3. Epidemiology   4. Hooke
5. Fungi   6. Leeuwenhoek   7. Spontaneous generation   8. Redi
9. Pasteur   10. Germ theory of disease   11. Koch
12. Metchnikoff   13. Semmelweiz   14. Reed   15. Ehrlich
16. Hypothesis   17. Human genome

Matching

1.a   2.c   3.b   4.j   5.g   6.d   7.i   8.h

9.b   10.d   11.g   12.d   13.f   14.e   15.a   16.i   17.h

Review Answers

1. **False** Kochs postulates were written for those organisms that can be purified on artificial culture medium. Since that is not possible with viruses and other obligate intracellular parasites, the postulates cannot be applied directly. (p. 13)

2. **True** Needham, a British clergyman, still believed that microbes arose spontaneously. By boiling and sealing flasks of broth, oxygen (air) needed for their "spontaneous generation" was not present. (p. 11)

3. **False** Leeuwenhoek, although credited with discovering the first microbes, did not construct the first microscope. In fact, his microscopes were considered very crude compared with his contemporaries. (p. 10)

4. **False** Fleming is credited with the identification of the first antibiotic, penicillin. (p.18) Chamberland developed porcelain filters used to remove bacteria. (p. 16)

5. **True** The United States Center for Disease Control is located in Atlanta, Georgia, and is the largest such agency in the world. (p. 5)

6. **True** Tyndall used airtight light boxes into which he placed flasks of boiled infusions. Once all the dust settled (which could be visualized through peep holes) the flasks would not become contaminated. (p. 12)

7. **e** Louis Pasteur's contributions were virtually endless. Because of his accomplishments, he is often known as the "Father of Modern Science". (pp. 11-12)

8. **c** Until Koch discovered the use of agar, microbes were usually grown on gelatin. Since most organisms degrade gelatin, this was not very suitable. By using agar, he was successful in purifying and thus proving the one organism one disease concept. (p. 13)

9. **d** Fungi are nucleated, unicellular and multicellular organisms that absorb their nutrients from the environment. Algae are photosynthetic, animals lack cell walls, and bacteria lack true nuclei. (pp. 3-4)

10. **d** Phycology is the study of algae, mycology is the study of molds, bacteriology is the study of bacteria, and virology is the study of viruses. (pp. 3-4)

CHAPTER 2

FUNDAMENTALS OF CHEMISTRY

For many students, one of the least popular subjects in science was chemistry. Perhaps it was due to a lack of adequate high school experience or perhaps it was due to just a bad experience. For many students it may have been the difficulty in understanding molecules and substances that could not be seen with the eye. Coupled with all of this is the problem that many students have difficulty understanding how chemistry relates to any of their other science courses such as microbiology and anatomy and physiology. However, because of the tremendous advances made in fields such as cell biology, physiology, and immunology, a basic understanding of general and organic chemistry is now truly essential. Without a knowledge of chemistry, the basis of the Gram stain, the structure and function of the cell membrane, the mechanisms of metabolism, and the processes involved in the immune response system could not be understood.

In the first part of this chapter, the basic building blocks of chemistry and the mechanisms of chemical bonding are presented. This is followed by a short discussion of the properties and features of water and other important inorganic substances including acids, bases, and salts. The last part of the chapter is devoted to a presentation of the complex organic molecules specifically, carbohydrates, lipids, proteins, and nucleic acids. Regardless of the type of background, this chapter will provide an excellent review of all the basic concepts of chemistry needed for an understanding of the chemical principles related to the field of microbiology.

STUDY OUTLINE

I.   **Why Study Chemistry**
II. **Chemical Building Blocks and Chemical Bonds**
     A.   Chemical building blocks
     B.   Structure of atoms
     C.   Chemical bonds
     D.   Chemical reactions
III. **Water And Solutions**
     A.   Water
     B.   Solutions and colloids
     C.   Acids, bases, and pH
IV. **Complex Molecules**
     A.   Basic characteristics
     B.   Carbohydrates
     C.   Lipids
     D.   Proteins
     E.   Nucleotides and nucleic acids

REVIEW NOTES

A.   **Why is knowledge of basic chemistry necessary to understanding microbiology?**

This is a common question asked by virtually all students. It would seem that chemistry should be studied only by chemists. That may have been true decades ago when basic structures of organisms were not understood, but today we need to know more about how those structures actually function. That involves chemistry. To understand how the cell membrane functions, it is necessary to know something about the molecules that make up the structure. To understand how our immune system works, it is necessary to know something about the proteins that form the antibodies. Without a basic knowledge of chemistry that is impossible. In most cases, only a moderate amount of information is required to understand even the most complex mechanisms and structures.

B.   **What terms describe the organization of matter and what elements are found in living organisms?**

There are only a few basic terms that need to be understood. Elements such as those listed in Table 2.3 are made up of invisible atoms which, in turn, consist of even smaller particles, namely, protons, neutrons, and electrons. The protons determine the atomic number of the atoms and therefore the nature of the element. They are found in the nucleus along with the neutrons which together make up the atomic weight. The electrons are in energy levels or orbits found outside the nucleus and play a major role in how the atom combines with others. The number of protons which are positive in charge always equals the number of electrons which are negative in charge unless the atom has combined or reacted with other atoms. The first orbit is filled with two electrons, the second is filled with eight, as is the third orbit. Atoms whose outer electron shells are nearly full or nearly empty can either gain or lose a few electrons and thus react with other atoms.

C. What are the properties of chemical bonds and chemical reactions?

The loss or gain of electrons to arrive at completely full or empty orbits creates electrically charged atoms called ions. (See Table 2.2.) This allows atoms to form chemical bonds, in this case, called ionic bonds. Atoms that can accept or donate an equal number of electrons can also form bonds which are called covalent bonds. A third type of bond, the hydrogen or weak bond, is formed by an attraction of atomic particles rather than a donation or sharing of electrons. All three types of bonds hold the small and even the large organic molecules together.

Chemical reactions allow for the breaking or forming of these chemical bonds and usually involve the release (exergonic) or the utilization (endergonic) of energy.

D. What properties of water, solutions, colloidal dispersions, acids, and bases make them important in living things?

Water is essential to all living organisms. Because it is a polar molecule (**see Figure 2.5, below**), it acts as the perfect solvent or mixer.

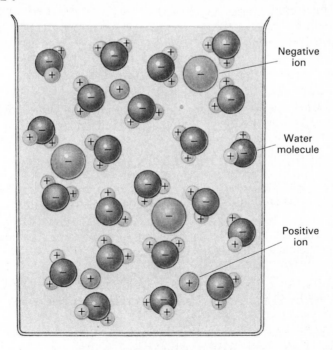

Figure 2.5: The polarity of water.

11

Water releases both a hydrogen ion and a hydroxyl ion so is neither an acid nor a base. Those molecules that release hydrogen ions can increase the acidity of a solution, and those that release hydroxyl ions can decrease the acidity. Chemists have devised the concept of pH to help measure the acidity of solutions. This is especially useful since living organisms can survive in only very narrow pH ranges.

E.   **What is organic chemistry and what are the major functional groups of organic molecules?**

Molecules that contain carbon are organic chemicals whereas those that don't are inorganic. Organic molecules frequently have important reactive or functional groups which are characteristic of that molecule. (**See Figure 2.8 below.**)

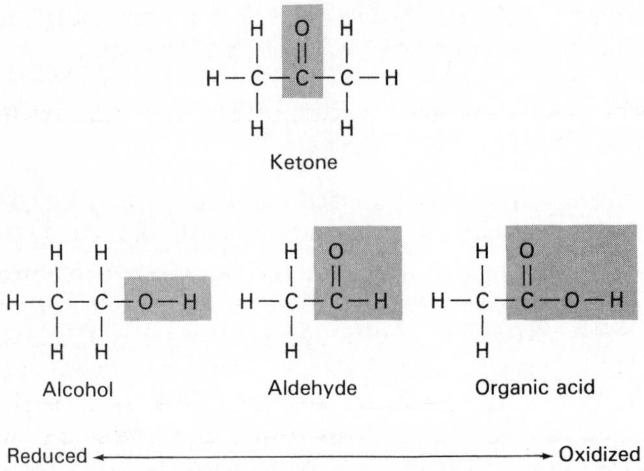

Figure 2.8: Functional groups.

F.   **How do the structures and properties of carbohydrates contribute to their roles in living things?**

Carbohydrates act as the primary energy molecules for cells and consist of carbon chains with various attached functional groups. Monosaccharides such as glucose are simple sugars and are the easiest to metabolize. Disaccharides such as sucrose and lactose must first be split in half before they can be eaten. Polysaccharides are long chains of monosaccharides bonded in a chain and are used as storage sugars.

G.   **How do the structures and properties of proteins, including enzymes, contribute to their roles in living things?**

Proteins act as structural and catalytic molecules for the cell. They are made of long chains of amino acids that can form into globular structures similar to a telephone cord that is coiled

12

up by continuously twisting in opposite directions (See Figure 2.18). Because of their functional amine and carboxyl groups, they are electrically charged as well. This allows them to form into important molecules such as membrane proteins, enzymes, and flagella. Because of their functional groups, they are, however, difficult to metabolize as food molecules and are not readily used for energy.

H. **How do the structures and properties of simple lipids, compound lipids, and steroids contribute to their role in living things?**

Lipids are long-chained and/or large cyclic organic molecules that are polarized, that is, they have electrically charged and uncharged ends (See Figure 2.14). This allows them to function as effective insulating and storage molecules and as membranes, since their polarity allows them to orient automatically in water similar to a fishing bobber. The cyclic molecules can act as important membrane stabilizers and as chemical messengers.

I. **How do the structures and properties of nucleotides contribute to their role in living things?**

Nucleotides, because of their size and shape, act as energy carriers and as carriers of genetic information. The phosphate groups of adenosine triphosphate hold a large amount of energy and allow it to act as the energy currency molecule for the cell. The DNA nucleotides of adenine, thymine, guanine, and cytosine form double chains in the form of a helix and act to carry coded genetic information based on the sequences of the nucleotides within the chain (See Figure 2.21). The RNA nucleotides of adenine, uracil, guanine, and cytosine form a single chain and act to decode the genetic information for the cell. The information coded by the nucleic acids allows for the formation of the wide variety of proteins needed by the cell.

LEARNING ACTIVITIES

Complete each of the following items by supplying the appropriate word or phrase.

1. The smallest chemical unit of matter is the _____.

2. The atomic number of an element is equal to the number of its _____.

3. A cation is a _____ charged ion.

4. Atoms of a particular element that contain different numbers of neutrons are called _____.

5. Bonds that form as a result of the attraction between ions of opposite charges would be _____ bonds.

6. Since catabolic reactions release energy, they can be called _____ reactions.

7. Water molecules possess a surface tension as a result of _____.

8. The medium in which substances are dissolved is the _____.

9. A substance which acts as a proton acceptor or a hydroxyl ion donor would be a _____.

10. A functional group that does not contain oxygen but does possess nitrogen would be an _____ _____.

11. Glucose represents an example of one of the three types of carbohydrates, namely a _____.

12. The bond that links two monosaccharides together to form a disaccharide is a _____ bond.

13. A type of lipid found in cell membranes which possesses phosphate groups would be a _____.

14. A type of lipid that exhibits four ring structures would be a _____.

15. Amino acids are unique molecules in that they always possess the _____ atom along with carbon, hydrogen and oxygen.

16. When proteins fold into a helix they exhibit the _____ structure.

17. A protein catalyst would be called an _____.

18. Nucleotides consist of three parts, namely, _____, _____, and _____.

19. A nitrogenous base found with RNA but not with DNA would be _____.

20. Another difference between DNA and RNA is that DNA is _____ _____.

21. When water is added to a reactant to form simpler products, the reaction is called a _____.

22. A deoxyribose sugar will have one less _____ than a ribose sugar.

Match the appropriate category for each of the molecules listed below.

Key Choices

a. Monosaccharide
b. Disaccharide
c. Polysaccharide
d. Fat
e. Amino acid

f. Phospholipid
g. Protein
h. Nucleic acid
i. Nucleotide
j. Steroid

_____ 1. Deoxyribose

__A__ 2. Glucose

__b__ 3. Sucrose

__C__ 4. Starch

_____ 5. Adenosine diphosphate

_____ 6. DNA

_____ 7. Triacylglycerol

_____ 8. Keratin

_____ 9. ATP

_____ 10. Cholesterol

Match the following terms with the descriptions listed below.

Key Choices:

a. Sodium
b. Carbon
c. Electron
d. Cation
e. Anion

f. Neutron
g. Proton
h. Mole
i. Atomic weight
j. Isotope

__g__ 11. Positively charged atomic particle.

__f__ 12. Negatively charged atomic particle.

__c__ 13. Negatively charged ion.

__s__ 14. An atom which forms covalent bonds with other atoms in a reaction.

_____ 15. Weight of a substance in grams equal to the sum of the atomic weights of atoms in a molecule of the substance.

15

_____ 16. Atoms of a particular element that contain different numbers of neutrons.

REVIEW QUESTIONS

True/False (Mark T for True, F for False)

_____ 1. Water is actually a polar compound.

_____ 2. An atomic particle that has mass but lacks an electrical charge is a neutron.

_____ 3. Covalent bonds are strong bonds formed by the exchange of electrons resulting in the formation of cations and anions.

_____ 4. Glucose and fructose are isomers because they have the same structure but differ in the number of atoms they possess.

_____ 5. Amino acids always possess amino, carboxyl, and variable (R) groups.

Multiple Choice

_____ 6. Which of the following is not true concerning water.
a. acts as a polar compound
b. has a low specific heat
c. forms thin layers due to its surface tension.
d. forms hydrogen bonds

_____ 7. Disruption of protein structure by changes in pH and/or high temperatures is called:
a. denaturation   c. hydrolysis
b. hybridization   d. complementation

_____ 8. The rule of octets refers to:
a. formation of isotopes.
b. chemically stable orbits.
c. the ratio of protons to neutrons.
d. the number of electron shells an atom possesses.

_____ 9. Lipids can be characterized by:
a. being soluble in water.
b. possessing large numbers of carbon and oxygen atoms and few hydrogen atoms.
c. forming amino acids on hydrolysis.
d. possessing ester bonds.

_____ 10. Select the most correct statement.
   a. Amino acids are all acidic and nonpolar molecules.
   b. Nucleic acids consist of long polymers of nucleotides.
   c. Fatty acids are always unsaturated, while phospholipids are always saturated molecules.
   d. ATP is an energy-carrying catalytic protein.
   e. All of the above are correct statements.

_____ 11. Which of the following does not relate to the nucleic acid, DNA.
   a. Hydrogen bonding between nucleotide bases.
   b. Complementary base pairing.
   c. A five carbon ribose sugar.
   d. Guanine, cytosine, thymine, and adenine.
   e. All of the above relate to DNA.

ANSWER KEY

Fill-in-the-Blank

1. Atom 2. Protons 3. Positive 4. Isotopes 5. Ionic
6. Exergonic 7. Hydrogen bonding 8. Solvent 9. Base
10. Amino group 11. Monosaccharide 12. Glycosidic bond
13. Phospholipid 14. Sterols 15. Nitrogen 16. Secondary
17. Enzyme 18. Nitrogenous base, 5 carbon sugar, phosphates
19. Uracil 20. Double stranded 21. Hydrolysis 22. Oxygen

Matching

1.a 2.a 3.b 4.c 5.i 6.h 7.d 8.g 9.i 10.j

11.g 12.c 13.e 14.b 15.h 16.j

Review Answers

1. **True** Water molecules have a region of partial positive charge associated with the hydrogen atoms and a region of partial negative charge associated with the oxygen atoms. This allows water to act as an excellent dissolving medium. (p. 32; Figure 2.4)

2. **True** Neutrons possess a relative mass of one and occupy the nucleus, but do not possess a charge. Protons, however, possess a positive charge and have a mass of one. (p. 28)

3. **False** Covalent bonds are strong bonds that actually share electrons rather than exchange them. Ionic bonds exchange electrons and form ions. (pp. 30-31; Figure 2.3)

4. **False** Glucose and fructose are isomers in that they have the same molecular formulas but differ in their structure and properties. (p. 36; Figure 2.9)

5. **True** Amino acids are the building blocks of protein and consist of one amino group ($-NH_2$), one carboxyl group ($-COOH$), and one R or variable group which varies with each amino acid. (p. 41; Figure 2.16)

6. **b** Water actually has a high specific heat in that it can absorb or release large quantities of heat energy with little temperature change. (p. 33)

7. **a** Temperatures above 50°C can denature or disrupt protein structures as well as highly acidic or basic conditions. Hydrolysis means to break bonds using water, hybridization means to cross breed, and complementation means the linking of nucleotide base pairs. (p. 42)

8. **b** When atoms are chemically stable, their outer-most shell (for many large atoms) contains 8 electrons. Atoms that have nearly full (6 or 7 electrons) or nearly empty (1 or 2 electrons) outer shells can form ions. (p. 29)

9. **d** The bond associated with lipids is the ester bond. With fats, it connects three fatty acids to a glycerol molecule. (p. 38; Figure 2.13)

10. **b** Nucleic acids such as DNA and RNA consist of long chains of the nucleotides, thymine, uracil, adenine, guanine, and cytosine. Amino acids can be acidic or basic, polar or nonpolar. Fatty acids and phospholipids can be saturated or unsaturated. ATP is an energy-carrying nucleotide. (p. 44)

11. **c** DNA contains a five carbon deoxyribose sugar. RNA contains a five carbon ribose sugar. (p. 44; Table 2.5)

CHAPTER 3

MICROSCOPY AND STAINING

The science of microbiology began in a series of small but
very critical steps. The first was the development of the microscope which probably occurred in the 16th century. The second was
the utilization of the microscope to actually see these invisible
microbes. That was accomplished by Anton van Leeuwenhoek in 1676.
The next step was the development of stains to enable scientists to
adequately see and study them. The subsequent steps involved
proving that microbes arise from living organisms and not spontaneously, and then proving that microbes actually cause disease.
Still, the fundamental link was the microscope.

The first part of the chapter presents the basic principles of
light which are often overlooked even though these principles are
necessary to understand how microscopes provide us with visual
images. Once these principles are understood, the techniques of
dark-field, phase contrast, and electron microscopy can be easily
explained. The last part of the chapter is devoted to the
principles behind the most important stains used in microbiology
such as the Gram, acid-fast, and endospore stains. Hopefully all
of this material will be further explained and demonstrated in the
laboratory.

STUDY OUTLINE

I.   **Development of Microscopy**
    A.   Early history
    B.   Anton van Leeuwenhoek
II. **Principles of Microscopy**
    A.   Metric units
    B.   Properties of light: wavelength and resolution
    C.   Properties of light: light and objects
III. **Light Microscopy**
    A.   Basic features
    B.   The compound light microscope
    C.   Darkfield microscopy
    D.   Phase-contrast microscopy
    E.   Fluorescence microscopy
    F.   Differential interference contrast microscopy
IV. **Electron Microscopy**
    A.   Basic features
    B.   Transmission electron microscope (TEM)
    C.   Scanning electron microscope (SEM)
V.   **Techniques of Light Microscopy**
    A.   Preparation of specimens for the light microscope
    B.   Principles of staining
    C.   The Gram stain
    D.   Special stains

REVIEW NOTES

A.   How is the evolution of microscopy instruments related to progress in microbiology?

Although the activities of bacteria and other microbes were apparent for thousands of years, the organisms were completely invisible to the eye. Therefore, the existence of these organisms was understandably delayed until the advent of the microscope. However, the first microscopes built were probably used to observe objects already visible such as fly wings, ants, and plant leaves. Fortunately, the inquisitive mind of Anton van Leeuwenhoek developed a microscope to see and record the presence of these tiny animalcules. As new microscopic techniques were discovered, more and more information about the structure and nature of these organisms was obtained.

B.   Which metric units are most useful for the measurement of microbes?

Since bacteria are too small to be seen or to be measured using the standard English system, a system of measurement called the metric system was developed (See Table 3.1). In this system, a meter is approximately equal to a yard and a millimeter is approximately equal to 1/16 of an inch. Both of these units can be visualized with the naked eye. The unit used to measure bacteria is the micrometer, which is 1/1000 of a millimeter, and the unit used to measure viruses is the nanometer, which is 1/1000 of a micrometer. Both of these units are too small for us to see but they do provide us with units to measure these very tiny organisms.

C.   **What are the relationships among wavelength, resolution, numerical aperture, and total magnification?**

Most of our microscopes use light as the energy source. Since light must pass between two objects for them to be resolved, it becomes the limiting factor (**see Figure 3.6 below**).

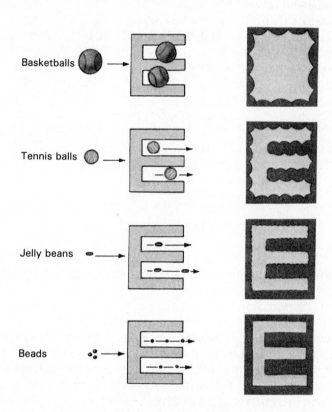

Basketballs

Tennis balls

Jelly beans

Beads

Figure 3.6: Wavelength and resolution analogy.

If the wavelength is too long to pass through them they will appear as one object. Therefore the smaller the objects, the harder it is to see or resolve them. Numerical aperture relates to the cone of light that enters the lens. The wider it is, the greater the resolving power of the lens.

D.   **How are the following properties of light related to microbiology: transmission, absorption, fluorescence, luminescence, phosphorescence, reflection, refraction, and diffraction?**

21

Light exhibits a wide variety of properties as it passes through substances of different densities. each of these properties affects how we can see objects viewed under a microscope. **See Figure 3.7 below** for an excellent summary of these properties.

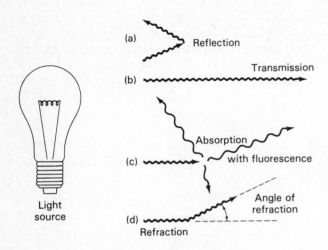

Figure 3.7: Properties of light.

Remember that the use of immersion oil in the oil immersion technique prevents the loss of light due to refraction.

E. **What is the function of each part of a compound microscope?**

All compound light microscopes have virtually the same components. Most of these are self-explanatory; however, be sure to understand the functions of the condenser lens, which converges light but does not magnify the object, and the iris diaphragm, which controls the amount of light.

F. **What are the special uses and adaptations of brightfield, darkfield, phase contrast, differential interference contrast, and fluorescence (UV) microscopes?**

The brightfield microscope, which is the most commonly used laboratory microscope, is characterized by a bright background and dark object and is best used to visualize stained specimens. The darkfield microscope exhibits a dark background and bright objects and is best used with live specimens. The phase contrast provides a contrast between the object and background and can be used with both stained and unstained specimens. Differential interference microscopes are very similar to phase contrast but provide a nearly three-dimensional image and are usually used with live specimens. Fluorescence microscopy involves the use of fluorescent dyes and is best used in clinical laboratories for diagnosis.

G.   What are the principles of transmission and scanning electron microscopy? How do the advantages and limitations of electron microscopy compare with those of light microscopy?

Electron microscopes differ from light microscopes in five important ways. The electron scope is more expensive and difficult to use; it uses electrons as an energy source; it uses magnets to focus the electrons; and it always requires a viewing screen or photographic plate to view the object. The use of electrons with very short wavelengths provides greater magnification and resolution of the objects; however, specimen preparation is far more difficult.

The TEM blasts electrons at the specimen and provides a flat image of the object while the SEM scans the object with a thin beam, thus generating a 3-D image.

H.   What techniques are used to prepare and heighten contrast in specimens to be viewed with a light microscope?

Wet mounts are the simplest techniques to use and only require a slide, cover slip, and a specimen. The organisms are live and unstained and generally hard to see. By air drying, heat fixing, and staining them with dyes, the organisms can be visualized.

I.   What are the uses of the common types of microbial stains?

The acid-fast stain allows detection of mycobacteria which may cause leprosy or tuberculosis. The Schaeffer-Fulton spore stain allows visualization of resistant endospores produced by members of the genus Bacillus and Clostridium. Negative stains use acidic dyes that do not penetrate or damage the cells and thus can be used to visualize fragile organisms. Flagellar stains are used to exhibit the flagella of motile organisms.

J.   What are the functions and results of each of the steps in the Gram staining procedure?

The Gram stain is the most important stain in microbiology (See Figure 3.28). The first step is the use of crystal violet dye, which stains all organisms purple. The second step is the use of iodine, which acts as a mordant to fix the dye into the cell wall layers. All organisms are still purple. The third step is the alcohol decolorizer step which removes the dye from the gram negative cells. Alcohol dissolves the lipids found in the outer layer of gram negative cells. The final step is the addition of safranine, which provides a contrasting color to the colorless gram negative cells.

23

LEARNING ACTIVITIES

Complete each of the following items by supplying the appropriate word or phrase.

1. Bacteria can be measured in metric units called _____.

2. Viruses should be measured in metric units called _____.

3. The ability to see two items as separate items is known as _____.

4. The widest cone of light that can enter an objective lens would be called _____.

5. When light passes through an object it would be known as _____.

6. _____ refers to the reemission of absorbed light as light of longer wavelengths.

7. Keeping the specimen in focus even when objective lenses are changed means the lenses are _____.

8. A compound microscope with a single eyepiece is said to be _____.

9. Fluorescent dyes used to stain organisms that cause tuberculosis and syphilis are called _____.

10. A microscopic adjustment knob that changes the distance between the lens and the specimen very slowly would be the _____.

11. The spraying of a heavy metal such as gold at an angle in the preparation of electron microscope specimens is known as _____ _____.

12. Positively charged dyes that are commonly used in bacteriological stains are _____ dyes.

13. Iodine acts as a ___Mordant___ in the Gram stain to fix the basic dye into the cell wall.

14. Stains which color the background and not the organism are called ___negative___ stains.
    safranin

24

15. Certain bacteria form highly resistant structures called _Endospores_ from free-living, reproducing cells called _vegetate_ cells.

16. In microscopy, the thickness of a specimen that is in focus at any one time is called the _____ of _____.

17. A new microscope that uses a thin wire probe to trace the surface of atoms which then creates electron clouds is called a _____ _____ or _____ microscope.

Match the following terms with the descriptions listed below.

Key Choices:

a. Transmission
b. Absorption
c. Fluorescence
d. Luminescence
e. Phosphorescence
f. Reflection
g. Refraction
h. Diffraction

_b_ 1. When light rays neither bounce off nor pass through and are taken up by an object.

_g_ 2. Bending of light as it passes from one medium to another of different density.

_e_ 3. Reemission of light waves after irradiation ceases.

_f_ 4. The rebound of light off an object which then provides its color.

_c_ 5. Reemission of absorbed light only during irradiation.

Match the following microscopes with their descriptions listed below.

Key Choices:

a. Brightfield
b. Darkfield
c. Phase contrast
d. Fluorescence
e. Differential interference
f. Transmission electron
g. Scanning electron

_____ 6. Condenser reflects light off of specimen rather than allowing light to pass directly through.

_____ 7. Possesses special condensers to accentuate small differences in the refractive index of specimens.

_____ 8. Used to reveal internal structures of thin slices of cells at resolutions down to 1 nanometer.

_____ 9. Uses UV light to excite molecules within specimen or with dyes that adhere to specimen.

_____ 10. Most commonly used laboratory microscope for students.

REVIEW QUESTIONS

True/False (Mark T for True, F for False)

__F__ 1. Leeuwenhoek constructed the first two-lens or compound microscope which enabled him to observe the tiny "animalcules."

__F__ 2. Nomarski interference creates a microscopic image where the background is dark and the organism appears bright.

__T__ 3. Immersion oil increases resolution but does not actually increase magnification.

__T__ 4. The acid-fast stain allows for the observation and detection of mycobacteria.

_____ 5. Ocular micrometers are used to measure ^both^ viruses when using the brightfield microscope whereas a stage micrometer is used to measure bacteria.

Multiple Choice

__C__ 6. The total magnification of a microscope with the low power lens (10X) and ocular lens (15X) in position would be: Ocular x objective
          a. 25   c. 150
          b. 15   d. 1500

_____ 7. A major difference between the SEM and the TEM is that the SEM:
          a. can resolve objects smaller than 20 nanometers.
          b. requires less of a vacuum system than the TEM.
          c. can create three-dimensional images.
          d. does not require the use of any metal coating of the specimen.

__a__ 8. Heat fixation accomplishes all the following except:
          a. helps the dye to penetrate the cells.
          b. kills the bacteria on the slide.
          c. decreases distortion of the cells prior to the addition of stains.
          d. fixes the organisms to the slide.
          e. All of the above are true.

26

_A_ 9.   The condenser lens of a microscope:
        a.   increases the magnification.
        b.   generally can magnify an object ten times.
        c.   increases the light refraction.
        (d.) converges light beams onto the specimen.

_B_ 10. If a bacterium measures 0.3 micrometer , it would measure
         how many angstroms?
         a.   300   c.   3000
         b.   30   d.   3

_b_ 11. Select the most correct statement about the Gram stain.
         a.   Iodine acts as the decolorizer in the Gram stain. no
         b.   Gram-positive cells retain the crystal violet dye and
              the gram-negative cells lose the dye following the
              decolorizer step. yes
         c.   To obtain the best Gram stain reaction, your cultures
              should be at least 48 hours old. no
         d.   Bacteria become gram-variable because they can change
              from a gram-negative type of cell wall to a tough
              gram-positive type of cell wall. no
         e.   All of the above are incorrect. _

ANSWER KEY

Fill-in-the-Blank

1. Micrometers 2. Nanometers 3. Resolution
4. Numerical aperture 5. Transmission 6. Luminescence
7. Parfocal 8. Monocular 9. Fluorochromes 10. Fine adjustment
11. Shadow casting 12. Cationic 13. Mordant 14. Negative
15. Endospores; vegetative 16. Depth of field
17. Scanning probe; tunneling

Matching

1.b 2.g 3.e 4.f 5.c

6.b 7.c 8.f 9.d 10.a

Review Answers

1.   **False**   Leeuwenhoek was the first to see microbes but he
constructed a very crude single lens microscope.   400 X
(p. 52; Figure 3.1)

2.   **False** Nomarski interference contrast produces images similar
to phase contrast where the image is virtually 3-D.
(p. 61; Figure 3.17 and Table 3.2)

3.   **True** Immersion oil has the same refractive index of glass and thus allows light to pass from the specimen to the lenses without refraction and loss of resolution. (p. 57; Figure 3.10)

4.   **True** Mycobacteria have complex lipid components in their cell walls which allows them to resist the acid alcohol used in the acid-fast stain. (p. 67; Figure 3.27)

5.   **False** Ocular and stage micrometers are both used to measure bacteria but since viruses cannot be seen with a light microscope, they are not used for their measurement. (p. 59)

6.   **c**   Total magnification of a microscope can be determined by multiplying the power of the ocular times the power of the objective lens. (p. 59)

7.   **c**   The SEM, because it involves the detection of scattered electrons, can provide a 3-D image. Specimens must still be coated, must be generally larger than 20 nanometers, and must still be observed under a complete vacuum. (p. 65; Table 3.2)

8.   **c**   Actually, heat fixation can often distort an object. To observe the best morphology, a negative stain which does not require heat fixing should be used. (p. 66)

9.   **d**   A condenser lens does not magnify an object but does converge all the light beams directly on the specimen being observed. (p. 58)

10. **c**   Since there are 1000 nanometers in a micrometer and 10,000 angstroms in a micrometer, 0.3 micrometers would be equivalent to 3000 angstroms. (p. 53; Table 3.1)

11. **b**   Alcohol or acetone-alcohol acts as the decolorizer; cultures should be less than 24 hours old to have the best Gram stain reactions; bacteria become variable because of changes in their cell walls due to age but they do not change from a gram-negative type of cell wall to a gram-positive type of cell wall. (p. 68; Figure 3.28)

CHAPTER 4

CHARACTERISTICS OF PROKARYOTIC AND EUKARYOTIC CELLS

Now that you have gained an understanding of the chemistry of cells and have learned about the microscopes and stains we use, you are ready to learn about the structure and function of the cells themselves. Some of this information may be covered in other courses such as anatomy and physiology and biology, so perhaps this will be an opportunity to refresh the memory.

The chapter begins with a review of the structure and chemistry of prokaryotic cells such as found with the bacteria and rickett- sia. This information will become invaluable as you learn how antibiotics affect the microbes, how the Gram stain colors cells differently, and how some microbes become resistant to various antibiotics and disinfectants. The next part of the chapter provides an analysis of the structures and functions of eukaryotic cells such as those found in plant and animal tissues. Notice that a number of the structures found in prokaryotic cells are quite similar to those found in eukaryotic cells. The final part of the chapter provides an overview of all those mechanisms that operate to move material in and out of a cell. Note again that many of these same mechanisms are found in both prokaryotic and eukaryotic cells.

STUDY OUTLINE

I.   **Basic Cell Types**
     A. Prokaryotic cells
     B. Eukaryotic cells
II. **Prokaryotic Cells**
     A. Size, shape, and arrangement
     B. Overview of structure
     C. Cell wall
     D. Cell membrane
     E. Internal structure
     F. External structure
III. **Eukaryotic Cells**
     A. Overview of structure
     B. Cell membrane
     C. Internal structure
     D. External structure
IV. **Movement of Substances Across Membranes**
     A. Basic characteristics
     B. Simple diffusion
     C. Facilitated diffusion
     D. Osmosis
     E. Active transport
     F. Endocytosis and exocytosis

REVIEW NOTES

A. What are the characteristics of eukaryotic and prokaryotic cells?

Prokaryotic and eukaryotic cells have a number of similarities and differences between them. Look at Table 4.1 for a complete list but concentrate on the major ones first. These are the similarities in the cell membranes, ribosomes, and cytoplasm. Prokaryotes, however, have very complex cell walls not found with the eukaryotes, but they lack the complex organelles found in eukaryotes (mitochondria and golgi bodies) which are as large or larger than the prokaryotic cell. Prokaryotes also lack a nuclear membrane that is found with eukaryotic cells.

B. How do prokaryotic cells differ in size, shape, and arrangement?

Prokaryotes, specifically the bacteria, have three basic shapes, namely, cocci, bacilli, and spirilli. Only the cocci have the greatest variety of arrangements (see Figure 4.1). Although we often see the bacteria in only two dimensions, try to visualize them as three dimensional rods and cocci (see Figure 4.2).

C. How are structure and function related in bacterial cell walls and cell membranes?

Bacterial cell walls surround the entire cell and are very complex. Notice that the cell wall construction directly relates to the staining properties of the cell. Those that are Gram-positive have a thick wall made almost entirely of peptidoglycan. The thick layer allows the cell to retain the purple dye. Gram-negative cells, on the other hand, have a thin inner layer of peptidoglycan and a relatively thick outer membrane of lipopolysaccharide material. This combination ensures that as the alcohol is added, any crystal violet dye that is adhering to these cells will be quickly removed since alcohol dissolves the lipid layer.

The lipopolysaccharide layer is also important because it can represent a toxic layer. Since it is a part of the cell, it is called the endotoxin.

Acid-fast bacteria also have a complex wall which includes lipids and waxy substances. These waxes greatly resist staining and especially decolorization steps. The presence of the waxes also explains why these cells do not readily mix in water when making a smear. (See Table 4.2 below)

TABLE 4.2 Characteristics of the cell walls of gram-positive, gram-negative, and acid-fast bacteria

Characteristic	Gram-Positive Bacteria	Gram-Negative Bacteria	Acid-Fast Bacteria
Peptidoglycan	Thick layer	Thin layer	Relatively small amount
Lipids	Very little present	Lipopolysaccharide	Mycolic acid and other waxes and glycolipids
Outer membrane	Absent	Present	Absent
Periplasmic space	Absent	Present	Absent
Cell shape	Always rigid	Rigid or flexible	Rigid or flexible
Effects of enzyme digestion	Protoplast	Spheroplast	Difficult to digest
Sensitivity to dyes and antibiotics	Most sensitive	Moderately sensitive	Least sensitive

In contrast to the porous, protective, and lifeless bacterial cell wall, the cell membrane is very active and functional. The lipids of the membrane act primarily as a barrier while the proteins act as the active functional units, determining what moves in or out of the cell and providing sites for cellular respiration. Interestingly, the membrane is much like a chocolate-chip cookie, with the dough acting as the lipids (boring, tasteless barrier) and the chocolate chips acting as the proteins (active, tasty and functional units) (see Figure 4.7).

D. How are structure and function related in other bacterial components?

Bacterial cells have only a few other internal and external parts. The internal parts include the ribosomes that act as sites of protein synthesis and the inclusion granules that act as storage units for the cell. The nuclear material, which contains the genes, lacks a nuclear membrane and consists of only one circular chromosome. This is similar to a large ball of string that is connected end to end. The external structures include the incredibly thin but long flagella used for motion and the glycocalyx which acts as a slime layer or capsule external to the cell wall. A thick capsule can protect the cell by resisting engulfment by white blood cells. Finally, some bacteria have short, stiff bristles called pili that can be used for surface contact or for contacting other cells in the process of gene transmission (**see Figure 4.3 below**).

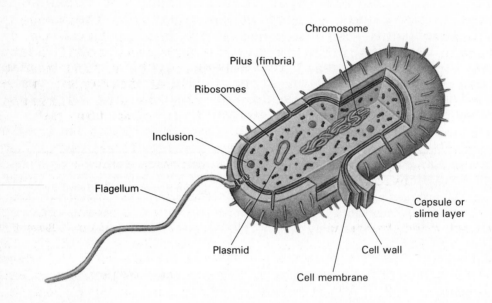

Figure 4.3: Prokaryotic structure.

E. How are structure and function related in eukaryotic cell membranes?

Eukaryotic cells are generally larger in structure and far more complex than prokaryotes (see Figure 4.17). The cell membrane of these cells is very similar to that found in prokaryotes except that they often possess sterols, probably to aid in stabilizing or strengthening the membrane. Since eukaryotes have specialized organelles for respiration and secretion, the cell membrane functions primarily to move material in and out of the cell.

F. How are structure and function related in other eukaryotic components?

Eukaryotes possess specialized organelles such as mitochondria which are used to produce energy, golgi bodies which are used for packaging and secreting substances, and endoplasmic reticula which are used for internal transport of materials. The nucleus is very complex and possesses pairs of chromosomes rather than the single chromosomes found with prokaryotes. This allows eukaryotes a greater degree of variation and genetic change. Numerous other structures may be found within the cytoplasm, each with a particular function fitted to its structure (see Figure 4.17).

External structures include flagella and cilia, both of which are considerably larger and more complex than the flagella of bacteria. See Figure 4.22 for a good comparison of sizes.

G. How do passive transport processes function and why are they important? How does active transport function and why is it important?

All living cells must move food material into their cytoplasm and move wastes out. This essential process is taken up by the cell membrane and specifically by the proteins found scattered within the lipid layer. Passive, non-energy requiring mechanisms such as osmosis, diffusion, and facilitated diffusion move the majority of materials. All operate on the basis of diffusion, in other words, if food is more abundant outside the cell (high concentration), it moves inside (low concentration) and vice versa for waste materials. Osmosis operates as a pulling force because the membrane is very selective. It, in effect, acts like a sponge. Facilitated diffusion involves the action of carrier membrane proteins that merely select which substances to bring in or take out of the cell (see Figure 4.25). Note that osmosis can create environmental conditions that can greatly affect cells. **See Figure 4.28 below** to visualize these effects, but keep in mind that a condition that adversely affects one microbe may in fact be favorable to another. For example, freshwater organisms placed in marine waters would be adversely affected as well as marine organisms placed in fresh water.

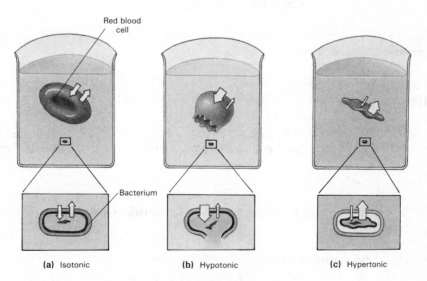

(a) Isotonic   (b) Hypotonic   (c) Hypertonic

Figure 4.28: Osmotic solutions.

H. **How do exocytosis and endocytosis occur and why are they important?**

Endocytosis and exocytosis are two energy-requiring mechanisms that move material through cell membranes, however, these only occur in eukaryotic cells. Phagocytosis is the most important example of endocytosis and involves the engulfing or "eating" of large particles such as cells or bacteria. Exocytosis is the reverse mechanism where a cell releases digested particles following phagocytosis (see Figure 4.30).

LEARNING ACTIVITIES

Complete each of the following items by supplying the appropriate word or phrase.

1. Organisms that lack a defined nucleus would be called _____Pro karyotic_____ cells.

2. _____Plleomorphic_____ bacteria vary widely in their cellular form even within a single culture.

3. A tough, interlinked component of the cell wall of bacteria that provides rigidity is the _____Peptidoglycan_____.

4. The two repeating subunits of the peptidoglycan layer are _____N-Acetylglucosamine_____ and _____N-Acetyl muramic acid_____.

5. A toxic component of the outer layer of Gram-negative bacteria is _____Lipid A_____.

6. If an organism loses its cell wall, the resultant structure is called a _____Protoplast_____.

7. An antibiotic that affects the formation of the cell wall is _____Penicillin_____.

8. Nonpolar hydrocarbon ends of fatty acids are said to be _____Hydrophobic_____ in respect to water. _Don't like H2O_

9. Accumulations of polyphosphate granules with the cytoplasm of bacteria are called _____Metachromatic_____ or _____volutin_____ granules.

10. Members of the genus <u>Bacillus</u> and <u>Clostridium</u> produce resistant structures called _____Endospores_____.

11. Flagella distributed all over the surface of bacteria are called _____Peritrichous_____ flagella.

12. Spirochaetes may possess a series of internal filaments used for motion called ___AXIAL___ ___filaments___.

13. Any substance containing polysaccharides found external to the cell wall is called the ___glyco CALYX___.

14. Molecules that add rigidity to all membranes of eukaryotes are ___Sterols___.

15. ___Histones___ are proteins bound to the DNA of eukaryotic cells.

16. Mitochondria are characterized by extensive inner membrane folds called ___cristae___.

17. A eukaryotic organelle which contains digestive enzymes is a ___Lysosome___.

18. Movement of molecules from a region of higher to lower concentration by means of a carrier protein is called ___Facilitated diffusion___.

19. A membrane that allows the passage of only a selected group of substances is said to be ___selectively permeable___.

20. Cells immersed in an ___isotonic___ solution would exhibit no change in their cell volume.

21. Cocci that form into long chains may be identified by the prefix ___streptococci___. ooooo

Match the following terms with the descriptions listed below.

Key Choices:

a. Fimbriae
b. Glycocalyx
c. Periplasmic space
d. Exosporium
e. Endotoxin
f. N-acetylglucosamine
g. Tubulin
h. Pseudopodia

___A___ 1. Short, hollow bristles used for attachment to surfaces.

___g___ 2. A protein that forms microtubules found in eukaryotic flagella.

___f___ 3. A component of peptidoglycan.

___d___ 4. The outer covering of an endospore. is Exosporium

___H___ 5. Temporary projections of cytoplasm associated with movement in eukaryotic cells.

_e_ 6. Endotoxin (toxic layer)
The outer lipid bilayer of most Gram-negative cells.

Match the following terms with the descriptions listed below.

Key Choices:

a. Osmosis   e. Active transport
b. Diffusion   f. Isotonic
c. Hypotonic   g. Facilitated diffusion
d. Hypertonic   h. Endocytosis

_d_ 7. A solution that will cause cells to shrink.

_A_ 8. The movement of water through a semipermeable membrane.

_H_ 9. The physical ingestion of particles as by phagocytosis.

_e_ 10. A movement of molecules through a membrane by means of carrier proteins expending energy.

_b_ 11. The mechanism by which perfume spreads around a room.

REVIEW QUESTIONS

True/False (Mark T for True, F for False)

_T_ 1. All prokaryotic and eukaryotic cells possess nuclear material and a fluid mosaic plasma membrane.

_F_ 2. Spiral bacteria have no arrangement commonly exhibit multicellular forms such as chains, tetrads, and grapelike clusters.

_F_ 3. Gram-negative bacteria possess a complex lipopoly- have saccharide cell wall complete with a lipid bilayer, but lack the peptidoglycan component found in Gram-positive a organisms.

_t_ 4. The nuclear membranes of all eukaryotic cells contain pores to allow for communication between the nucleus and the cytoplasm.

_F_ 5. Both mitochondria and golgi bodies found in eukaryotes contain DNA and can replicate independently.

_T_ 6. Hypertonic solutions may cause cells to lose water and shrink.

_F_ 7. When gram-negative bacteria are killed, their toxicity is also destroyed. but by releasing endotoxin

Multiple Choice

_C_   8.   Which of the following would not be consistent with eukaryotic organisms.
   a. membrane bound organelles
   b. presence of histones
   c. cell membranes lacking sterols
   d. paired chromosomes

_b_   9.   The peptidoglycan layer of the cell wall:
   a. consists of lipopolysaccharide.
   b. is formed from repeating molecules of gluNAc and murNAc.
   c. represents the lipid bilayer.
   d. lacks teichoic acids.

_A_   10.   Dipicolinic acid is commonly associated with:
   a. endospore coats.
   b. lipopolysaccharide of Gram-negative bacteria.
   c. peptidoglycan layer of Gram-positive bacteria.
   d. mesosomes.
   e. None of the above is true.

_b_   11.   Bacterial flagella:
   a. attach to the cell wall via the teichoic acids and calcium.
   b. form a hook after leaving the cell.
   c. are about the same size as eukaryotic flagella.
   d. are composed of lipopolysaccharide units called flagellin.

ANSWER KEY

Fill-in-the-Blank

1. Prokaryotic 2. Pleomorphic 3. Peptidoglycan
4. N-acetylglucosamine; N-acetylmuramic acid 5. Lipid A
6. Protoplast 7. Penicillin 8. Hydrophobic
9. Metachromatic; volutin 10. Endospores 11. Peritrichous
12. Axial filaments 13. Glycocalyx 14. Sterols 15. Histones
16. Cristae 17. Lysosome 18. Facilitated diffusion
19. Selectively permeable 20. Isotonic 21. Strepto-

Matching

1.a 2.g 3.f 4.d 5.h 6.e

7.d 8.a 9.h 10.e 11.b

Review Answers

1. **True**   These characteristics are shared by both types of organisms. (p. 76; Table 4.1)

2. **False**   Spiral organisms generally exist only as single cells. Bacilli often form chains and cocci can exhibit a variety of shapes. (p. 77; Figure 4.1)

3. **False**   Gram-negative bacteria do possess a lipopolysaccharide layer but also possess a thin peptidoglycan layer.
(p.80; Table 4.2)

4. **True**   Eukaryotic cells have nuclear pores which allow RNA molecules to leave the nucleus and participate in protein synthesis. (p. 91; Figure 4.18)

5. **False**   Both mitochondria and chloroplasts contain DNA and can replicate. Golgi bodies lack DNA and are incapable of division. (p. 92)

6. **True** Hypertonic solutions such as brine water may cause cells to shrink due to the osmotic "pulling" effect of the excess salt outside the cells. (pp. 99-100)

7. **b** Gram-negative bacteria have an endotoxin which is released when they die. Therefore killing them may only increase the toxicity. (p. 80)

8. **c** Eukaryotic cell membranes do contain sterols whereas procaryotic membranes do not. Mycoplasmas do possess sterols in their membranes. (p. 76; Table 4.1)

9. **b** The peptidoglycan layer forms a chain link fence type of multiple layers of gluNAc and murNAc. The other items represent components or features of the lipopolysaccharide layer associated with the outer membrane of Gram-negative bacteria.
(p. 78; Figure 4.4)

10. **a** Endospore coats contain layers of peptidoglycan, calcium, and dipicolinic acid. (p. 86)

11. **b** Bacterial flagella are composed of protein subunits called flagellin; attach to the cell wall and membrane; are considerably smaller than eukaryotic cell flagella; and form a hook immediately after leaving the cell. (p. 87; Figure 4.12)

CHAPTER 5

ESSENTIAL CONCEPTS OF METABOLISM

A major characteristic of all living organisms is their ability to produce energy from foods and to use this energy to synthesize needed materials. When Louis Pasteur made the discovery that microbes were involved in the making of wine, many other unknown phenomena such as the souring of milk and the spoilage of meats were explained. This knowledge led to the discoveries of the processes of fermentation and respiration and to the cycles of glycolysis, Krebs, and electron transport. These in turn led to the use of microbes for industrial production of hundreds of chemicals, byproducts, and enzymes.

The first part of this chapter provides an overview of the properties and characteristics of enzymes that catalyze all the chemical reactions in living organisms. This is followed by a presentation of anaerobic metabolism, specifically the properties and mechanisms of glycolysis and fermentation and aerobic metabolism with the Krebs cycle and electron transport system. The last part of the chapter explains the mechanisms involved in noncarbohydrate metabolism, photosynthesis, and other similar systems seen in environmentally important microbes.

STUDY OUTLINE

REVIEW NOTES

A. How do the following terms relate to metabolism: autotrophy, heterotrophy, oxidation, reduction, photoautotrophy, photo-heterotrophy, chemoautotrophy, chemoheterotrophy, glycolysis, fermentation, aerobic metabolism, and biosynthetic processes?

Metabolism is the sum total of all chemical reactions in a cell which includes catabolism or energy-producing reactions and anabolism or energy-requiring reactions. This energy is found in the electrons of all foods used by organisms. By removing them (oxidation) and adding them to other compounds (reduction) they can obtain the energy they need.

Various mechanisms are used to obtain this energy. Organisms can use simple chemicals or the sun to gain energy and are called autotrophs, or organisms can use ready-made organic molecules and are then called heterotrophs (see Figure 5.2 below).

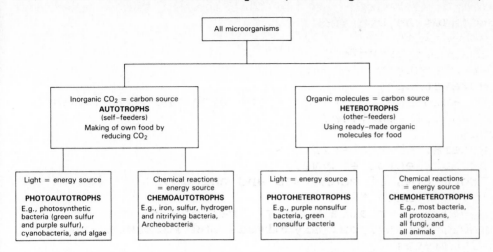

Figure 5.2: Kinds of energy-capturing metabolism

As this chapter will describe, various pathways and cycles are used such as fermentation and respiration in order to extract the energy found in foods.

B. **What are the characteristics of enzymes and how do those characteristics contribute to their function?**

Enzymes are amazing organic catalysts. They are highly specific and exhibit a "lock and key" type of mechanism as they react with a substrate (see Figure 5.5). Keep in mind that they are made of protein and are therefore genetically coded in the DNA.

Some enzymes require helpers or coenzymes to react with substrates. These coenzymes are simple chemicals in comparison to the enzyme itself.

At certain times, it is necessary to control the enzyme activity. One way is to compete with or block the active site, sort of like a chastity belt. Another is to distort the active site by reacting with an inhibitor (noncompetitive inhibition). A final way is to turn the enzyme production system off at the DNA or gene level.

C. **What are the main steps and significance of glycolysis and fermentation?**

Fermentation is a very simple catabolic system used by a great number of organisms to produce their energy. Fermentors produce very little energy per food (net 2 ATP), which means they are very inefficient. Therefore they must eat a lot of food to fulfill their energy needs. As a result, they also produce a lot of wastes which are organic acids, gases, and/or alcohols. The only cycle they usually use is glycolysis. Notice that the beginning substrate is usually glucose and that the end product is pyruvic acid. All fermentors use this common pathway. Once they reach

41

pyruvic acid they each go to different final organic end products as shown in **Figure 5.11 below**. Notice that many of these final end products are quite useful in industry.

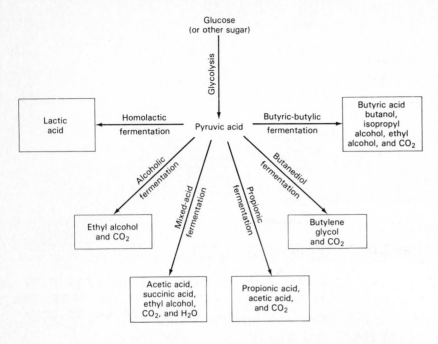

Figure 5.11: Final end products.

D. **What are the main steps and significance of the Krebs cycle?**

While fermentation is a simple process utilizing only glycolysis, aerobic respiration is more complex involving three processes, namely, glycolysis, the Krebs cycle, and the electron transport system. Because more systems are used, aerobes produce much more energy per food (36-38 ATP) and are therefore more efficient. They don't need to eat as much food and they also produce fewer waste products. Aerobes, for the most part, are not as useful in industry as anaerobes.

As mentioned, glycolysis is used initially to produce pyruvic acid which is then converted to acetyl COA. This then enters the Krebs cycle (see Figure 5.16). The cycle produces only a few, but important, products. A small amount of energy is produced, carbon dioxide is released, and a lot of high energy electrons are extracted for use in the third cycle.

E. **What are the roles of electron transport and oxidative phosphorylation in energy capture?**

Once electrons are extracted from molecules within the Krebs cycle by the help of the coenzymes, NAD and FAD, they can enter the electron transport system (see Figure 5.17). This system allows for the production of tremendous amounts of ATP or energy currency. The system operates similar to water running down hill.

42

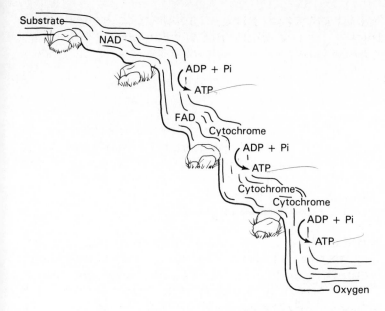

Figure 5.18: Electron transport system.

Once the electrons have been "relieved" of their energy they must be accepted by a final electron acceptor. The acceptor in aerobic respiration is oxygen, hence the term oxidative phosphorylation. (**See Figure 5.18 above**)

F. How do microorganisms metabolize fats and proteins for energy?

All foods contain energy. Even doughnuts, which are made of starch and sugar, contain enormous quantities of energy. This energy is, in fact, found in the electrons of each sugar molecule. All that living organisms have to do is use a system which will allow them to extract the energy from those electrons and convert it to ATP.

Although sugars are the best and easiest to use for energy, even fats and proteins can be used. Fats can be broken down by beta-oxidation systems to form acetyl-COA which can enter the Krebs cycle. Proteins are more difficult to metabolize and must first be deaminated and then altered before they can enter the cycles.

G. What are the main steps and significance of photosynthesis in microbes?

All photosynthetic microbes are harmless since they are autotrophic; however, they are very significant in the environment. All of these organisms possess a variety of pigments that are used in photosynthesis. Some, especially those found around thermal pools, can really enhance the beauty of the environment.

Photosynthetic bacteria differ from plants in four ways. First, their chlorophyll absorbs in a different range; second, they often use organic compounds to reduce CO_2; third, many release H_2S or sulfuric acid; and fourth, they do not release oxygen as a

byproduct. Most photosynthetic bacteria are also strict anaerobes.
The cycles used by bacteria in photosynthesis are, however,
essentially the same as those found in green plants.

H.   **How do photoheterotrophy and chemoautotrophy differ?**

Photoheterotrophs and chemoautotrophs are all harmless but
important environmental organisms. Photoheterotrophs use the sun's
energy and organic carbon molecules while chemoautotrophs use very
simple inorganic chemicals for their carbon and energy source. The
latter have recently been found around deep sea volcanic vents.

I.   **How do bacteria carry out biosynthetic activities?**

Once organisms have produced energy, they can utilize it for
the biosynthesis of building material for the cell wall and cell
membrane, for cellular movement, and for growth. Some pathways
called amphibolic pathways can produce energy and also construct
building blocks at the same time. Figures 5.25 and 5.26 illustrate
some of the building blocks that are commonly produced.

J.   **How do bacteria use energy for membrane transport and for
movement?**

Membrane transport involves permease enzymes which allows
substances to be concentrated inside the cell or perhaps for
storage. Movement occurs by expending energy to move flagella or
axial filaments. See chapter 4 and especially Figure 4.12. Note
how the rings of a flagellum are especially designed to rotate
within the wall and membrane when energy is applied.

LEARNING ACTIVITIES

Complete each of the following items by supplying the appropriate
word or phrase.

1.   Reactions that require energy to synthesize complex molecules
     from simpler ones are known as _____.

2.   When a substance gains electrons, it is said to be
     _____.

3.   Organisms that can use carbon dioxide to synthesize organic
     molecules are _____Autotroph_____.

4.   Organisms which obtain energy from sunlight and which use
     ready-made organic molecules are _____.

5.   Enzymes which act within the cells that produce them are
     _____.

6. The substance that an enzyme reacts with is the _Enzyme Substrate Complex_.

7. The substance that is formed once the enzyme has changed it is called the _Holoenzyme Product_

8. A nonprotein substance that acts with the enzyme to catalyze a reaction is a _Coenzyme_.

9. A _____ adds water to break large molecules into two smaller molecules.

10. An enzyme that joins two molecules together is a _Ligase_.

11. If an inhibitor binds at a site other than the active site which then alters the shape of the active site, it is called _Allosteric_ _Inhibition_.

12. The addition of a phosphate group often from ATP to another molecule is called _____.

13. The common end product of the glycolysis cycle is _Pyruvic_ _Acid_.

14. The cycle that metabolizes two carbon units to CO_2 and H_2O is the _____.

15. The molecule that enters the Krebs cycle from glycolysis is _____.

16. The process by which fatty acids are catabolized is known as _Transition_.

17. Before amino acids can enter into any metabolic system they must first be _____.

18. During photosynthesis, energy from electrons is used to split water to form protons, electrons, and oxygen in a process called _____.

19. Bacteria that can oxidize inorganic substances for energy are known as _____.

20. Enzymes that extend through the membrane to aid in active transport of molecules are called _____.

Match the following terms with the descriptions listed below.

Key Choices:

a. Holoenzyme   d. Allosteric site
b. Apoenzyme   e. Active site
c. NAD   f. FAD

45

_b_ 1. The protein portion of the enzyme.

_NM)_ 2. A coenzyme which is made from the vitamin niacin.

_d_ 3. The end product can bind here to inactivate the enzyme by altering its shape.

_F_ 4. Uses the vitamin riboflavin as its active portion.

Match the following terms with the descriptions listed below.

Key Choices:

a. Glycolysis   e. Alcoholic fermentation
b. Krebs cycle   f. Oxidative phosphorylation
c. ETS   g. Substrate-level phosphorylation
d. Chemiosmosis   h. Homolactic acid fermentation

_A_ 5. Operates in the presence or absence of oxygen.

_A_ 6. Pyruvic acid is a major end product of this system.

_e_ 7. Pyruvate is converted to lactate in this system.

____ 8. Energy in GTP is converted directly to ATP.

____ 9. Represented by a series of coenzymes.

____ 10. The process of energy capture where protons are pumped out of the matrix to ultimately make ATP.

____ 11. This system converts pyruvate to acetaldehyde which is then converted to ethyl alcohol.

REVIEW QUESTIONS

True/False (Mark T for True, F for False)

____ 1. Autotrophic organisms rarely cause human disease.

____ 2. Enzymes increase the amount of activation energy needed to initiate a reaction.

____ 3. The glycolysis cycle is required by anaerobes in the process of fermentation but it is not used by aerobic microbes in the process of respiration.

____ 4. NAD and FAD operate in the Krebs cycle to remove high energy electrons and hydrogen protons for transfer to the electron transport system.

46

_____ 5. Because anaerobes produce only 2 ATP per food molecule (versus 38 ATP in aerobic systems), they grow very slowly and probably require many days to reach any appreciable population size.

_____ 6. Photosynthesis in green plants differs from that in bacteria in that the bacteria generally do not produce oxygen as a byproduct.

Multiple Choice

_e_ 7. Which of the following is true concerning enzymes?
a. Enzymes are generally only used once in a reaction.
b. Enzymes always end with the suffix -ole.
c. Enzymes consist of large lipopolysaccharide molecules.
d. Enzymes form an enzyme-product complex before completing their reaction.
e. None of the above is true.

_d_ 8. The Krebs cycle is characterized by all the following except:
a. substrate-level energy capture.
b. oxidation of carbon.
c. removal of electrons by coenzymes.
d. formation of pyruvate as an end product.

_____ 9. In the "dark" reaction of photosynthesis:
a. electrons in chlorophyll are activated by sunlight.
b. photolysis occurs.
c. carbon dioxide is reduced to form glucose.
d. NADP is reduced.
e. all of the above do not occur.

_____ 10. An amphibolic pathway:
a. generates energy in photolithotrophs.
b. can capture energy or synthesize substances needed by the cell.
c. is required to metabolize fats to fatty acids.
d. provides the surfactants needed for movement.

ANSWER KEY

Fill-in-the-Blank

1. Anabolism 2. Reduced 3. Autotroph 4. Photoheterotroph
5. Endoenzyme 6. Substrate 7. Product 8. Coenzyme (or cofactor)
9. Hydrolase 10. Ligase 11. Noncompetitive inhibition
12. Phosphorylation 13. Pyruvic acid 14. Krebs cycle
15. Acetyl COA 16. Beta oxidation 17. Deaminated 18. Photolysis
19. Chemolithotrophic 20. Permeases

Matching

1.b 2.c 3.d 4.f

5.a 6.a 7.h 8.b 9.c 10.d 11.e

Review Answers

1.   **True**   Autotrophic or self-feeding metabolism especially, photosynthesis, is an important means of energy capture. Since they are free-living, they usually do not cause disease. (p. 110)

2.   **False**   Enzymes decrease the activation energy needed for reactions thus facilitating them. (p. 112; Figure 5.4)

3.   **False**   Glycolysis is an essential metabolic cycle for both systems. (pp. 117-119; Figure 5.10)

4.   **True**   Both NAD and FAD are coenzymes which function to accept electrons from molecules in the Kreb cycle and transfer them to coenzymes of the ETS in the process of oxidative phosphorylation. (p. 123; Figure 5.17)

5.   **False**   Although anaerobes may grow more slowly than aerobes under optimum conditions for both, they still can reach maximum population densities in about the same time as aerobes. (p. 124)

6.   **True** Photosynthetic bacteria differ from green plants in that bacterial chlorophyll absorbs longer wavelengths of light than plant chlorophylls, they use hydrogen compounds other than water for reducing carbon dioxide, and they do not release oxygen as a byproduct. (p. 128)

7.   e   Enzymes are catalysts and can be used over and over in a reaction; they always end with an -ase; they are composed of protein; and they form an enzyme-substrate complex. (pp.112-117)

8.   d   The Krebs cycle forms CO_2 as one of its endproducts. Pyruvic acid is an end product of the glycolysis pathway. (pp.121-123)

9.   c   In the dark reaction, carbon dioxide is reduced by electrons from NADP to form glucose. The reduced coenzyme and ATP are formed by the light reactions. (pp. 128-129; Figure 5.24)

10. b   Since many intermediate compounds are involved in biosynthesis, the pathways can now be called amphibolic rather than catabolic. (p. 130)

CHAPTER 6

GROWTH AND CULTURING OF BACTERIA

Bacteria have an amazing ability to reproduce at astounding rates. In order to properly determine the numbers they generate, we have to use logarithmic scales. Just consider a small sample of milk or meat left on the counter overnight. This sample can be transformed from one virtually free of microbes to one consisting almost entirely of microbial parts and pieces!

This chapter focuses on the mechanisms and phases of bacterial growth and cell division and on how bacterial growth can be measured. It discusses the numerous physical and nutritional factors that affect growth. This section also helps to explain how organisms have had to adjust to every changing environment. Sporulation is presented as an important aspect of how some groups of organisms have adapted to rather harsh environmental conditions.

The last part of the chapter emphasizes the important methods used in pure culturing organisms and the methods and culture media used in growing these organisms. Much of this material will hopefully be emphasized by interesting laboratory experiments.

STUDY OUTLINE

REVIEW NOTES

A.   How is growth defined in bacteria?

Growth is defined as the orderly increase in the quantity of all the components of an organism. Since bacteria have a limited increase in size, their growth is measured by an increase in the number of cells.

B.   How does cell division occur in microorganisms?

Cell division in bacteria occurs primarily by binary fission (see Figure 6.1). That is accomplished by duplicating its components, forming a transverse septum, and splitting into two cells. Yeast cells and some bacteria may also divide by budding, in which a new cell forms from the surface of an existing cell.

C.   What are the phases of growth in a bacterial culture?

There are four major phases of growth, namely, the lag, log, stationary, and decline or death phase (**see Figure 6.2**). During the lag phase, the organisms are metabolically active but are not increasing in number. The log phase exhibits the rapid logarithmic growth pattern based on the organisms' generation time or the time needed to divide one cell into two at regular intervals. Note the formula on p. 137 for calculating the generation times. A simpler but less accurate formula for determining the number of bacteria that can reproduce in a period of time is 2^N where N = the total number of generations an organism has made in that given time. For example, if an organism can divide once every 30 minutes and is given 10 hours to divide, the total number of bacteria in the population would be 2^{20} or about 1,000,000 bacteria (assuming you have started with only one cell). Now consider a glass of milk left on the counter during the day and think of how many bacteria could be generated! Awe inspiring.

The third phase is the stationary phase which occurs because many of the organisms begin to die off due to the production of wastes, lack of food, and overcrowding. The decline phase occurs as the number of cells dying is far greater than those being reproduced. Once this phase is reached, the entire population may soon die off.

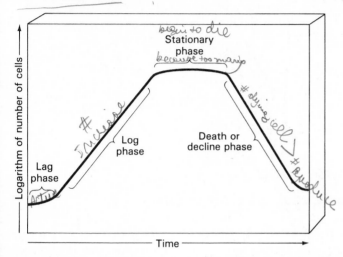

Figure 6.2: Phases of growth.

D. How is bacterial growth measured?

Methods to measure bacterial growth include the colony count, which is accomplished by transferring a known dilution of bacteria onto an agar plate and counting the colonies that arise. Other methods include direct microscopic counts, most probable numbers, filtration, turbidity, and others.

E. How do physical factors affect bacterial growth?

Various physical factors affect the growth of bacteria either through its affect on the cell's enzymes, on its metabolic systems, or on its structure such as the cell wall or membrane. However, because of the wide diversity of organisms, many have evolved to survive under almost any condition.

Acidity and alkalinity of the medium affect microbial growth and most organisms have an optimum pH range, with some preferring an acidic, some an alkaline, and some a neutral range. Temperature also affects organisms, with some preferring a high temperature, some a low temperature, and some a moderate one. See Figure 6.12 which illustrates the ranges that varying groups of organisms require.

It is important to remember that pathogens seem to prefer a neutral pH and a moderate temperature range.

goD

Oxygen also affects microbes, with some preferring aerobic conditions, some anaerobic conditions, some with a reduced amount of air, and some without any preference. Other factors that affect microbes include water and its osmotic effect and hydrostatic pressure.

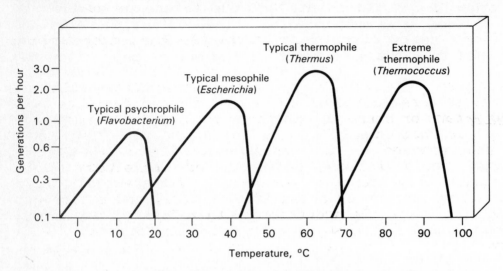

Figure 6.12: Growth rates and ranges based on temperature.

F.   How do nutritional factors affect bacterial growth?

Nutritional factors also greatly affect the growth of microbes. All require a carbon source supplied either organically (sugars) or inorganically (CO_2); a nitrogen source supplied either organically (amino acids) or organically (ammonia); and several other elements such as sulfur, phosphorous, and potassium. Vitamins are also required but some can be synthesized by the organism.

Organisms' nutritional requirements are determined by the type of enzymes they possess. Various bioassay techniques can be used to determine the needs they require. Organisms can adjust to their environment by producing exoenzymes accordingly, by making enzymes to metabolize another available nutrient, or by adjusting their metabolic activities.

G.   What occurs in sporulation and what is its significance?

Certain bacteria in the genus *Bacillus* and *Clostridium* produce highly resistant structures called endospores. The sporulation process occurs continuously to ensure survival under almost any condition (see Figure 6.15). When favorable conditions arise, the endospores undergo germination and the cells return to the normal or vegetative state.

H.   What methods are used to obtain a pure culture of an organism for study in the laboratory?

Various methods are used to grow and pure culture microbes. The most common method is the streak plate which involves spreading bacteria on a solid medium by means of an inoculating loop. The pour plate is a second method where organisms are pipetted into a dish and cooled liquid agar is added. The individual organisms are then free to develop into isolated colonies within the gel.

I.   **How are different nutritional requirements supplied by various media?**

Various culture media have been devised for microbial cultivation. Media may be constructed with known or defined contents such as glucose and water, or unknown, undefined (complex) contents such as beef extract or "hot dogs." Complex media are the most common to use since they provide all the elements needed for growth.

Specialized media include selective or inhibitory media designed to grow only certain organisms, differential media designed to distinguish one organism from another, and enrichment media designed to enhance the growth of certain "fussy" organisms.

Organisms can be maintained on stock cultures for routine work or as preserved cultures to store and save for long periods of time.

LEARNING ACTIVITIES

Complete each of the following items by supplying the appropriate word or phrase.

1.   A bacterial cell duplicates its components and divides into two cells by ____Binary____ ____Fission____.

2.   Organisms that have adapted to a medium thus allowing for rapid exponential growth are in the ____Log____ phase. growing in size

3.   If bacteria could divide together and have the exact same generation time they could exhibit ____STATIONARY____ ____phase____. (synchronous growth)

4.   During the decline phase, many cells exhibit unusual shapes which is caused by ____Involution____.

5.   A method used to measure bacterial growth that requires the use of a series of dilution tubes is ____serial dilution____.

6.   ____Turbidity____ or a cloudy appearance, is an indication of bacterial growth.

7.   Organisms that can tolerate very low pH conditions are called ____Acidophilic____.

8.   ____Thermophilic____ are organisms that prefer temperatures above 50°C.

9. The term _Facultative_ means that the organism can tolerate one environmental condition but still live in another.

10. An enzyme which can break H_2O_2 into H_2O and O_2 is _____.

11. Organisms that grow best in the presence of small amounts of oxygen are called _microaerophilic_.

12. Obligate anaerobes are killed by a highly reactive form of oxygen called _super oxide_.

13. When cells lose water and their membranes shrink away from the cell wall they are undergoing _Plasmolysis_.

14. Organisms that have many special nutritional needs are _Fastidious_.

15. Enzymes that are produced by the cell to operate outside the cell are _Extra cellular_ or _____.

16. The process of endospore formation is known as _Sporulation_.

17. The two bacterial genera that produce endospores are _Bacillus_ and _Clostridium_.

18. A method of pure culturing bacteria by means of an agar plate and an inoculating loop is the _Streak Plate_.

19. A culture medium which contains known specific kinds and amounts of chemicals is a _Defined Synthetic_ medium.

20. A medium that encourages the growth of some organisms but suppresses others is a _selective_ medium.

21. Isolated organisms can be maintained in a pure culture called _Stock culture_.

22. A preserved culture maintained to keep its characteristics as originally defined is a _Reference_ culture.

23. A specially calibrated counting chamber used with direct microscopic counts is called a _____ counter.

24. Trace elements such as copper and zinc often serve as _cofactors_ in enzymatic reactions.

25. Microbes such as _E. coli_ are able to manufacture vitamin _K_ in the human intestinal tract.

Match the following terms with the descriptions listed below.

Key Choices:

a. Lag phase
b. Log phase
c. Stationary phase
d. Decline phase

e. Generation time
f. Synchronous growth
g. Chemostat
h. Nonsynchronous growth

A 1. An adjustment period prior to rapid growth.

C 2. When cell division decreases to the point that new cells are produced at the same rate as old cells die.

g 3. A device that maintains logarithmic growth for extended periods of time. Chemostat.

d 4. The period during which many cells involute.

f 5. A hypothetical stair-step pattern of growth. Synchronous growth

Match the following terms with the descriptions listed below.

Key Choices:

a. Acidophile
b. Alkalinophile
c. Thermophile
d. Psychrophile

e. Obligate aerobe
f. Facultative anaerobe
g. Microaerophile
h. Halophile

d 6. Organisms living in the arctic ocean.

h 7. Organisms that prefer moderate to large quantities of salt.

e 8. An organism that must grow in the presence of oxygen.

b 9. Organisms that live in alkaline soils.

g 10. Organisms that tolerate only a small amount of oxygen.

Match the following terms with the descriptions listed below.

Key Choices:

a. Defined medium
b. Nondefined medium

c. Selective medium
d. Differential medium
e. Enrichment medium

_b_ 11. A medium that would contain agar, peptones, and beef extract.

_c_ 12. A medium to which antibiotics and/or dyes have been added.

_e_ 13. A medium such as blood agar.

_a_ 14. A medium that contains only glucose, salt, and water.

REVIEW QUESTIONS

True/False (Mark T for True, F for False)

_F_ 1. Although organisms may differ in the length of the lag phase, all organisms have the same generation time.

Explenantial
Logarymic _F_ 2. Generally microbes divide at an arithmetic rate.

_T_ 3. Most bacteria that cause disease are in the neutrophilic category. Have neutral PH.

_F_ 4. An obligate anaerobe means it prefers an anaerobic condition but can tolerate a small amount of air.

_T_ 5. All organisms require a carbon and nitrogen source along with trace elements such as sulfur and phosphorous.

_T_ 6. Nutritional complexity reflects a deficiency in biosynthetic enzymes.

Multiple Choice

_b_ 7. Select the most correct statement in relation to serial dilution.
a. Diluted samples are transferred to nutrient broth tubes.
b. The number of colonies on the plate is multiplied by the denominator of the dilution factor.
c. The test can accurately measure live and dead cells.
d. Countable plates should contain between 10 and 30 colonies.

_e_ 8. Which of the following methods is not used to determine bacterial numbers?
a. turbidity
b. most probable number
c. serial dilution
d. direct microscopic counts
e. All of the above are used.

_b_ 9.   Temperatures can control bacteria because:
        a. freezing will kill all bacteria.
        b. hot temperatures (above 80°C) will denature bacterial
            protein.
        c. refrigerator temperatures stop the growth of all
            bacteria.
        d. no bacteria can live above 180°F.

_c_ 10.   Endospores:
        a. are generally formed for protection and reproduction.
        b. are formed only when conditions become unfavorable.
        c. contain dipicolinic acid and calcium.
        d. contain laminated layers of peptidoglycan called the
            exosporium.

_a_ 11.   Select the most incorrect statement concerning culturing
        of bacteria.
        a. A synthetic medium consists of unidentifiable
            ingredients such as those found in beef extract.
        b. Agar is a solidifying agent that melts at about 96°C.
        c. The streak plate method uses agar plates and a wire
            inoculating loop.
        d. An enrichment medium contains ingredients such as
            blood which can enhance the growth of certain
            organisms.
        e. All of the above are true.

ANSWER KEY

Fill-in-the-Blank

1. Binary fission 2. Lag 3. Synchronous growth 4. Involution
5. Serial dilution 6. Turbidity 7. Acidophiles 8. Thermophiles
9. Facultative 10. Catalase 11. Microaerophiles 12. Superoxide
13. Plasmolysis 14. Fastidious 15. Extracellular 16. Sporulation
17. _Bacillus_, _Clostridium_ 18. Streak plate
19. Defined synthetic medium 20. Selective 21. Stock culture
22. Reference culture 23. Petroff-Hausser 24. Cofactors 25. K

Matching

1.a 2.c 3.g 4.d 5.f

6.d 7.h 8.e 9.b 10.g

11.b 12.c 13.e 14.a

Review Answers

1.   **False** Organisms vary not only in the length of the lag phase
but also in the log phase. Generation times for most bacteria is
20 minutes to 20 hours. (pp. 138-139)

2.   **False**   Bacteria all divide at an exponential or logarithmic rate due to binary fission. (p. 139)

3.   **True**   Neutrophiles have an optimum pH near neutrality which is about pH 5.4 to 8.5. Since the pH of the human body is within that range it is understandable that most pathogens would be in this group. (p. 145)

4.   **False**   The term obligate means that the organism must have the environmental condition specified.   Therefore an obligate anaerobe can only tolerate complete anaerobic conditions.
(pp. 145-146; Figure 6.13)

5.   **True**   All organisms do require all of these nutrients; however, they can be supplied in either an inorganic or organic form. (p. 149)

6.   **True**   Organisms with fewer enzymes have complex nutritional requirements because they lack the ability to synthesize many of the substances needed for growth. (p. 150)

7.   **b**   The technique of serial dilution involves transferring diluted samples to agar plates for counting; the test can only detect live cells; and the countable plates should be in the range of 30 to 300. (p. 141; Figure 6.5)

8.   **e**   Numerous methods can be used to measure bacterial growth such as all the methods listed in the question and even an additional method called filtration. (pp. 141-144)

9.   **b**   Organisms vary as to their temperature requirements. Psychrophiles prefer cold temperatures around 15°C, mesophiles prefer moderate temperatures around 25°C - 30°C and thermophiles prefer hot temperatures around 60°C.   Freezing inactivates but does not necessarily kill microbes; refrigerator temperatures do not affect psychrophiles; and some extreme thermophiles grow at or near boiling. (p. 145; Figure 6.12)

10. **c**   Endospores are formed only for protection; are often formed continuously; and contain layers of peptidoglycan that form the cortex of the spore coat. (pp. 151-153; Figure 6.15)

11. **a**   A synthetic medium is one that can be prepared in the laboratory from materials of precise or well-defined composition. (pp. 153-155)

CHAPTER 7

GENETICS I: GENE ACTION, GENE REGULATION, AND MUTATION

Ever since the discovery of the gene code and its genetic carrying capacity, DNA has been the object of intense interest and research. A molecule at first so simple in construction but yet so complex in its informational carrying capacity and its regulation, it has provided microbiologists hundreds of interesting avenues of study.

Because bacteria are relatively simple prokaryotic organisms, they have been the primary tool for studying this intriguing molecule. Dozens of microbiologists have received Nobel prizes for their outstanding work involving microbial genetics. As a result of their efforts and the work of many others we have begun to understand the intricacies of this molecule and have been able to apply much of the information gained to the benefit of all mankind.

The first part of this chapter is devoted to an overview of the gene and characteristics of the DNA molecule. Protein synthesis is then explained to illustrate how DNA is translated into functional and structural molecules. Gene regulation, which is extremely important for the cell, is described next. This is followed by a discussion of how genes can be altered and mutated and what effect this will have on the cell. As a final note, read the essay at the end of the chapter. It describes a fascinating new technique called the polymerase chain reaction (PCR) which allows scientists to rapidly produce small segments of DNA without the use of living cells. This technique is now being used to identify and study a great variety of genes found in human cells.

STUDY OUTLINE

I.   **Overview of Genetic Processes**
   A.   Basis of heredity
   B.   Nucleic acids in information storage and transfer
II. **Replication of DNA**
   A.   DNA arrangement
   B.   Replication forks
   C.   Mechanism of DNA polymerase
   D.   Semiconservative replication
III. **Protein Synthesis**
   A.   Transcription
   B.   Kinds of RNA
   C.   Translation
IV. **Regulation of Metabolism**
   A.   Significance of regulatory mechanisms
   B.   Categories of regulatory mechanisms
   C.   Feedback inhibition
   D.   Enzyme induction
   E.   Enzyme repression
V.   **Mutations**
   A.   Types of mutations and their effects
   B.   Phenotypic variations
   C.   Spontaneous versus induced mutations
   D.   Chemical mutagens
   E.   Radiation as a mutagen
   F.   Repair of DNA damage
   G.   The study of mutations
   H.   The Ames test

REVIEW NOTES

A.   How are genes, chromosomes, and mutations involved in heredity in prokaryotic organisms?

All information necessary for life is stored in an organisms genetic material, its DNA. Heredity involves the transmission of this information from an organism to its progeny. This is accomplished through the duplication and transfer of genes that consist of linear-coded sequences of DNA. The genes make up large genetic units called chromosomes. Prokaryotes carry one circular chromosome while eukaryotes possess numerous pairs of chromosomes. The genetic information is transmitted during binary fission in prokaryotes. Mutations are permanent alterations in the DNA that can be transmitted to the progeny and appear to account for much of the variations seen in organisms.

B.   How do nucleic acids store and transfer information?

Deoxyribonucleic acid (DNA) makes up the chromosomes of all cells. The information is stored within the DNA in a coded message

60

determined by the linear arrangement of the nucleotides adenine, guanine, cytosine, and thymine. This information is used to replicate the DNA in cell division and to provide information for protein synthesis. The mechanism of complementary base pairing ensures the accurate coding and decoding of the information.

C.   How is DNA replicated in prokaryotic cells?

The replication of DNA begins at specific points in the circular chromosome and proceeds in both directions. **See Figure 7.4** for an overall view of the replication process in prokaryotes.

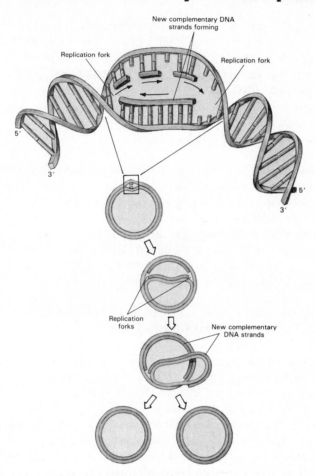

Figure 7.4: Prokaryote replication.

The process is semiconservative where each chromosome receives a strand of parent DNA and one newly synthesized strand of DNA.

D.   What are the major steps in protein synthesis?

The process of protein synthesis involves two major steps. In transcription, a messenger RNA molecule is transcribed or formed from a section of the DNA. In translation, this message is decoded by ribosomes into specific proteins such as enzymes or structural

units such as flagella. **See Figure 7.3 below** for a simplified view
of the information transmission from DNA to protein.

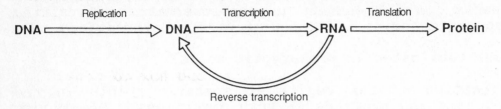

Figure 7.3: Overview of information transfer.

The entire process of protein synthesis is directed by three
types of RNA whose properties are outlined in Table 7.1. Look over
this table to understand the roles of each of these RNA molecules.

The necessary amino acids for this construction are brought to
the ribosome by transfer RNA molecules. Using an anticodon system
the tRNA's are matched to codons in the mRNA to insure an exact
sequencing of amino acids as directed by the mRNA.

This process is somewhat analogous to the *master computer* at
a college. This computer could represent the DNA of a cell.
Information requested about a single class from the campus could be
transcribed onto *floppy disks* which would act as the mRNA. These
disks could then be translated by placing them into a *disk drive* of
a small personal computer which would act as the ribosome. The
disk would be decoded onto *paper* or onto a *printed tape* which would
act as the finished product or protein produced by the cell. Now
look over Figure 7.10 to obtain an overview of the entire process
of protein synthesis from the cell's point of view.

**E.   How do mechanisms that regulate enzyme activity differ from
those that regulate gene expression?**

Regulatory mechanisms are essential to turn metabolic systems
on and off when necessary for the cell. This ensures that the cell
does not waste energy producing more than what it needs and ensures
the organism of essential materials.

Two basic regulatory mechanisms occur in prokaryote cells.
The first involves regulation of enzyme activity already in the
cell. This frequently involves feedback inhibition by the end
product. The second involves regulation of enzyme production by
affecting the action of the genes.

**F.   What happens in feedback inhibition, enzyme induction, and
enzyme repression?**

Feedback inhibition involves regulation of existing metabolic
enzymes associated with a biochemical pathway. This is accomplished by the end product binding to an enzyme inhibitor or
allosteric site which temporarily turns off the enzyme. Enzyme
induction involves the activation of genetic units responsible for
the production of a particular enzyme. See Figure 7.13 for a

review of this process. Enzyme repression involves the inhibition of genetic units responsible for enzyme production. These two systems appear to affect genetic units called operons and effectively control enzyme synthesis. The presence of the nutrient or absence of the end product appears to act as the conditions that lead to the gene expression.

G. What changes in DNA occur in mutations and how do they affect organisms?

Permanent changes in the DNA affect the genotype and are called mutations. These may or may not be expressed in the phenotype or what is actually seen. These mutations may be either point mutations which affect only a single nucleotide, or frameshift mutations which involve the insertion or deletion of nucleotides. See Table 7.3 for a review of these mutations.

H. How do spontaneous and induced mutations differ?

Spontaneous mutations occur in the absence of all known mutagenic agents and may be due to errors in base pairing during DNA replication. Induced mutations are produced by agents called mutagens which include base analogs, alkylating agents, radiation, and others. Some organisms have repair mechanisms that remove or replace the damaged sections of DNA. Remember that any change in the DNA can be permanent whereas changes in the RNA or protein synthetic products without changes in the DNA may not be permanent.

I. How do the fluctuation test, replica plating, and the Ames test make use of bacteria in studying mutations?

Microbes are excellent organisms for genetic studies because they reproduce in very large numbers, are easy to handle, are inexpensive to produce, and can express their genetic information and/or changes very quickly. As a result several tests have been developed to take advantage of these features. The fluctuation and replica plating tests (see Figures 7.21 and 7.22) use bacteria to demonstrate the spontaneous nature of mutations. The Ames test (see Figure 7.23) is based on the ability of bacteria to mutate and is used to screen chemicals for potential mutagenic properties.

LEARNING ACTIVITIES

Complete each of the following items by supplying the appropriate word or phrase.

1. Eukaryotic genes with different information at the same locus are called _____.

2. A permanent alteration in the DNA is a ___Mutation___.

63

3. When the DNA serves as a pattern in the formation of mRNA, it acts as a ___Template___.

4. The pairing of certain nucleotide bases in DNA to other nucleotides is called _____ _____ _____.

5. The process of mRNA formation from the DNA template is called ___Transcription___.

6. When RNA messages are decoded into proteins the process is known as _____.

7. DNA replication creates two moving strands called _____ _____.

8. Replication processes that form one strand of old or parent DNA and one of new DNA is called _____ _____.

9. An enzyme necessary to form mRNA from the DNA template is the _____ _____.

10. Each tRNA molecule has a three nucleotide base region that is complementary to a particular mRNA codon called an _____.

11. Codons such as UCC UAG GGU would be found in _____.

12. If an mRNA codon sequence such as UAA or UAG would appear during the reading cycle, it would act as the _____ _____.

13. An RNA molecule that brings the appropriate amino acid to the ribosome for packaging is the _____.

14. Enzymes that are synthesized continuously are said to be _____.

15. A series of closely associated genes which relate to the synthesis of specific structures and to their regulation is an _____.

16. A substance that binds to and inactivates the repressor associated with an operon is an _____.

17. When a needed metabolite such as glucose is in adequate supply within the cell and which then inhibits the activity of the operon, the effect is called _____ _____.

18. A genetic change that affects the DNA would be a ___Genotype___ change.

19. A mutation that results in a deletion or an insertion of single bases is a _____ mutation.

20. A nutritionally deficient mutant organism is an _Auxotrophic_ mutant.

21. A molecule similar to a nitrogenous base found in DNA is a _____ _____.

22. Mutagenic agents such as nitrous acid that can remove amino groups from nitrogenous bases are known as _____ agents.

23. Ultraviolet light creates thymine-thymine _____.

24. DNA repair that occurs in the presence of visible light can be called light repair or _____.

25. A test that is based on the ability of bacteria to mutate and which is used to screen chemicals for mutagenic properties is the _____ test.

26. A test commonly used to detect antibiotic resistant organisms is the _____ _____ test.

Match the following terms with the descriptions listed below.

Key Choices:

a. rRNA
b. mRNA
c. tRNA
d. RNA polymerase

e. DNA polymerase
f. Ligases
g. Exonucleases
h. Okazaki fragments

_____ 1. RNA molecules that have a cloverleaf shape with an attachment site for specific amino acids.

_____ 2. RNA that serves as a site for protein synthesis.

_____ 3. An enzyme that binds to a strand of DNA and initiates transcription.

_____ 4. Enzymes that remove segments of DNA.

_____ 5. An enzyme that can join DNA segments together.

Match the following terms with the descriptions listed below.

Key Choices:

a. Point mutations
b. Frameshift mutations
c. Constitutive enzymes
d. Inducible enzymes

e. Enzyme induction
f. Enzyme repression
g. Induced mutations
h. Spontaneous mutations

_____ 6. Insertion of acridine into a DNA helix can cause these types of mutations.

_____ 7. Enzymes that are produced only in response to a substrate.

_____ 8. A type of regulatory mechanism that is biosynthetic and uses energy.

_____ 9. Mutations that occur in the absence of any known agent.

_____ 10. Enzymes that are produced even when the substrate is absent.

REVIEW QUESTIONS

True/False (Mark T for True, F for False)

_____ 1. When two DNA strands combine by base pairing, they do so in a head-to-head or parallel fashion.

_____ 2. Although the first codon of the mRNA always codes for methionine it acts as the "start" signal.

_____ 3. Actually several ribosomes "read" an mRNA at the same time.

_____ 4. The operon theory was first proposed by Watson and Crick in 1951.

_F_ 5. All DNA mutations will ultimately result in the death of the cell. ⟩ change Genotype but dosn't kill the cell

Multiple Choice

_e_ 6. Feedback inhibition:
   a. allows the cell to conserve energy.
   b. acts to inhibit the active site by competitive inhibition.
   c. involves control by the end product.
   d. All of the above are correct.
   e. Only a and c are correct.

_c_ 7. In DNA replication:
   a. the old strands of DNA are first destroyed.
   b. a DNA-dependent RNA polymerase forms the parent strands.
   c. replication is semiconservative.
   d. DNA synthesis is continuous to ensure accurate synthesis.

___d___ 8.   Bacteria are ideal for studies on regulatory mechanisms
            because:
            a.   they reproduce at a very rapid rate.
            b.   they can be grown inexpensively.
            c.   they mutate at a fairly rapid rate.
            d.   All of the above are good reasons.

_____ 9.   In respect to the <u>lac</u> operon:
            a.   the absence of lactose results in the inhibition of
                 the repressor.
            b.   the repressor binds to the operator and turns on the
                 lac genes.
            c.   lactose acts as the inducer when it is present.
            d.   the inducer induces the repressor gene to turn off
                 the operator.

_____ 10. All of the following are true concerning mutagenic agents
            except:
            a.   they act at the genotypic level.
            b.   alkylating agents can add methyl groups to
                 nitrogenous bases.
            c.   once chemicals mutate the DNA, it cannot be repaired.
            d.   X-rays can easily break chemical bonds.

        11. Concept Question:

             How can a mutation in the DNA affect a cells polysacch-
             aride capsule?

                              ANSWER KEY

Fill-in-the-Blank

1.   Alleles   2. Mutation   3. Template
4.   Complementary base pairing 5. Transcription 6. Translation
7.   Replication forks   8. Semiconservative replication
9.   RNA polymerase 10. Anticodon   11. RNA   12. Terminator codons
13. tRNA   14. Constitutive   15. Operon   16. Inducer
17. Catabolite repression   18. Genotypic   19. Frameshift
20. Auxotrophic   21. Base analog   22. Deaminating   23. Dimers
24. Photoreactivation   25. Ames   26. Replica plating test

Matching

1.c 2.a 3.d 4.g 5.f

6.b 7.d 8.f 9.h 10.c

Review Answers

1.   **False** DNA strands combine by base pairing in a head-to-tail
or antiparallel fashion. (pp. 166-167; Figure 7.4)

                               67

2.   **True** Start and stop signals are part of the coded message in mRNA. The first RNA code is AUG which codes for methionine and acts as the start signal. (p. 170; Figure 7.8)

3.   **True** Several ribosomes called polyribosomes read the mRNA strand at the same time, attaching to the end of the mRNA that corresponds to the beginning of a protein. (p. 173; Figure 7.10)

4.   **False** The operon theory was first proposed by the French scientists, Jacob and Monod in 1961. Watson and Crick developed our knowledge of DNA and the genetic code. (pp. 175-176)

5.   **False** Mutations always change the genotype but it may not result in the death of the cell. If the mutation is a minor change in the DNA with no change in the amino acid specified by the mRNA there may not be any effect on the organism.
(pp. 178-183; Table 7.3)

6.   **e** Feedback inhibition, also called end product inhibition operates to control biochemical pathways by directly inhibiting the first enzyme in the path. The end product usually binds in the inhibitor site thus causing noncompetitive inhibition.
(p. 175; Figure 7.12)

7.   **c** DNA replication insures that each double helix consists of one old and one new strand. The old strands are not destroyed, the synthesis is discontinuous, and a DNA polymerase synthesizes the complementary strands. (pp. 166-169; Figure 7.5)

8.   **d** Bacteria are ideal for all types of genetic studies as outlined in this chapter; however, since they are not eukaryotic cells, an exact parallel cannot always be made. (p. 173)

9.   **c** The absence of lactose results in the synthesis of the lac repressor which then binds to the operator and prevents transcription of the genes. The inducer binds to and inactivates the repressor thus turning on the operon.
(pp. 175-177; Figure 7.13)

10. **c** Although mutagenic agents can alter the DNA, the damage can be repaired by a number of mechanisms including light and dark repair. (pp. 178-183; Figure 7.19)

11. Mutations affect the DNA which would then affect the mRNA and its ability to make specific proteins. If the affected protein is an enzyme that is required to construct the polysaccharide capsule, the result could be a defective capsule. (pp. 178-179)

CHAPTER 8

GENETICS II: TRANSFER OF GENETIC MATERIAL
AND
GENETIC ENGINEERING

One of the greatest discoveries in all of microbiology and perhaps in all of science was the finding that bacteria can actually transfer genetic material from one cell to another. Prior to this time, most microbiologists believed that microbes were unable to transmit genetic material and thus were incapable of making any significant changes. This initial discovery quickly opened the door to the exciting mechanisms of transformation, conjugation, and transduction.

Armed with the knowledge of these mechanisms, microbiologists were quick to apply them to agriculture, industry, and medicine. Using some enterprising modifications, the field of genetic engineering was born. Of all the current areas of microbiology, this one seems to offer the greatest promises for mankind as we begin to approach the twenty-first century.

The first part of this chapter provides a detailed analysis of the three main types of genetic transfer along with a discussion of extrachromosomal units called plasmids. The last part of the chapter is devoted to a discussion of the recent discoveries such as genetic fusion, gene amplification, and recombinant DNA technology that have made this field what it is today.

STUDY OUTLINE

REVIEW NOTES

A.   What is the nature and significance of gene transfer?

Gene transfer refers to the movement of genetic information between organisms and occurs in bacteria by transformation, transduction, and conjugation. This type of transfer is significant because it increases genetic diversity in a population and therefore insures that some of the members of the population will always survive environmental changes. These mechanisms were not thought to occur with bacteria until the discoveries by Griffith and others. See Table 8.2 for a summary of the three basic types of genetic transfer mechanisms.

B.   What is the mechanism and significance of transformation?

Bacterial transformation was discovered in 1928 by Griffith, who showed that live, non-lethal rough pneumococci and heat-killed previously lethal, smooth pneumococci could produce live, lethal pneumococci when introduced together into a mouse. This discovery was especially important because it demonstrated an ability of bacteria to change genetically. Following this discovery others,

70

including Avery and Watson and Crick, demonstrated that DNA was responsible for this genetic change and that the genetic material was coded in the DNA molecules.

Transformation involves the release of DNA fragments from donor cells and their uptake by recipient cells during specific stages in their growth cycle (see Figure 8.2). Competence factors are required for the DNA uptake so that endonucleases can cut the DNA into units and single strands of DNA can be incorporated into the recipient cell's DNA.

This mechanism is significant because it represents an important method of genetic transfer, it can be used in genetic manipulation, and it can be used to create recombinant DNA.

C.   What are the mechanisms and significance of transduction?

A second mechanism of genetic transfer is transduction. This involves the transfer of DNA from a donor cell to a recipient via a bacteriophage. When it was discovered that phages could be either virulent or temperate, the likelihood of this mechanism occurring became apparent. Transduction, therefore, requires two basic steps (see Figure 8.3 below).

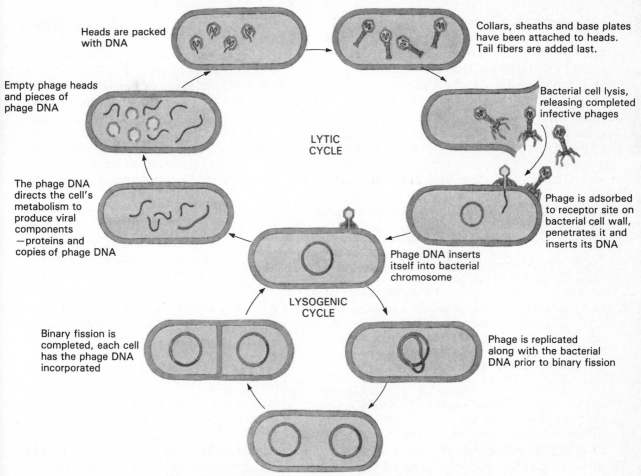

Figure 8.3: Lytic and lysogenic cycles of bacteriophages.

First, a virulent phage must infect and destroy a host cell's DNA and then cause lysis. In the process the virulent phage may pick up a few genes of the host cell and incorporate them into its genome. The second step involves the virulent phage infecting a new host (recipient) and instead of lysing the cell, it persists in the cell and becomes a temperate phage. The temperate phage, now called a prophage, can be incorporated with the recipient's DNA or it can exist as a plasmid. Bacterial cells containing the prophage are considered lysogenized since the virus could reenter the lytic phase.

Specialized transduction involves the transfer of only those host genes adjacent to the phage. Generalized transduction involves the transfer of any DNA fragment since the phage exists as a plasmid.

Transduction is significant because it represents another mechanism of genetic transfer and because it demonstrates that prophages and bacterial cells must have had a close evolutionary relationship. Furthermore, it suggests a viral origin of cancer and it provides a mechanism for studying gene linkage.

D. What is the mechanism and significance of conjugation?

Conjugation involves transfer of genetic material from a donor cell to a recipient cell via a conjugation bridge called a pilus. Donors are called "males" or F+ cells and recipients are called "females" or F- cells. See Figure 8.8 for an illustration of the basic process of conjugation. F+ cells can transfer extrachromosomal units called plasmids. If the plasmid becomes integrated into the host chromosome, the F+ cell, now called an Hfr, can transfer a large number of host genes. If the plasmid is incorporated and subsequently separated from the host chromosome, the cell becomes an F' cell and can transfer host genes that are attached to the plasmid. **See Table 8.1 below** for an excellent summary of the results of selected conjugations.

TABLE 8.1 Results of selected conjugations

Donor	Recipient	Molecule(s) Transferred	Product
F^+	F^-	F plasmid	F^+ cells
Hfr	F^-	Initiating segment of F plasmid and variable quantity of chromosomal DNA	F^- with variable quantity of chromosomal DNA
F'	F^-	F plasmid and some chromosomal genes it carries with it	F plasmid and some duplicated chromosomal genes

This mechanism is significant because it represents another method of genetic transfer and it increases the prospects of genetic diversity in bacteria. It helps provide another means of mapping bacterial chromosomes and indicates additional evolutionary stages of development.

E.   **What are the characteristics and actions of plasmids?**

Plasmids are circular, self-replicating, double-stranded extrachromosomal units of DNA that carry nonessential but often useful pieces of genetic information (see Figure 8.11). Some plasmids carry antibiotic resistant genes called R factors, some provide information to produce bacteriocins which can inhibit the growth of other bacteria, some are virulence plasmids that cause disease symptoms, and some can cause tumors in plants. Most, if not all, plasmids carry genetic information to produce the pili necessary for transfer of the plasmids.

F.   **How are the following techniques of genetic engineering used: (a) genetic fusion, (b) protoplast fusion, (c) gene amplification, (d) recombinant DNA, and (e) hybridomas?**

Genetic engineering is the manipulation of genetic material to alter the characteristics of an organism. In order to accomplish this, several techniques have been developed. Genetic fusion involves the alteration of genes within a single species of organism. This is accomplished by deleting small sections or relocating certain sections to create different characteristics. Protoplast fusion involves the combination of two, cell wall-less organisms with the resultant fusion of their DNA. This allows the mixing of favorable characteristics of two different organisms. Gene amplification involves the addition or multiplication of plasmids in organisms to increase the yield of products the cell can produce. This has increased the antibiotic production of certain organisms. Recombinant DNA technology involves the insertion of foreign DNA into the DNA of bacteria and usually involves plasmids. This technology has provided a mechanism to produce very large quantities of substances such as human insulin, interferon, and interleukins made only by eukaryotic cells. Hybridomas are genetic recombinations usually involving eukaryotic cells. This technique has allowed the production of large quantities of highly specific antibodies and other molecules.

G.   **Why are scientists concerned about uses of recombinant DNA?**

Recombinant DNA technology has carried with it advantages and disadvantages. The advantages included the ability to alter genes of any organism to produce large quantities of proteins that were very difficult to produce. This was contrasted with the prospect that the technology might create a virulent uncontrollable pathogen. Fortunately, this prospect seems very remote at the present time.

73

LEARNING ACTIVITIES

Complete each of the following items by supplying the appropriate word or phrase.

1. The first mechanism of gene transfer which was discovered by Griffith was _Transformation_.

2. A special protein that is released into a medium that appears to facilitate the uptake of DNA is _Competence Factor_.

3. Viruses that specifically infect bacteria are called _Bacteriophage_.

4. A virus that does not actually kill its host cell but instead incorporates into it is a _Temperate_ phage.

5. Extrachromosomal circular pieces of DNA are called _Plasmids_

6. The protein bridge which is used in conjugation is _Pilus_.

7. A microbial strain that can induce a very large number of genetic recombinations is a _HFR_ strain.

8. Plasmids that commonly carry genes for resistance to antibiotics are called _R Factors_.

9. Bacteriocinogens produced by several strains of E. coli are _Colicins_.

10. The fusion of two, cell wall-less bacteria is called _Protoplast Fusion_.

11. The addition of plasmids to microorganisms to increase the yield of useful substances is known as _____.

12. DNA that contains information from two different species of organisms is _Recombinant_ DNA.

13. A self-replicating carrier such as a phage or plasmid is a _____.

14. One important application of hybridomas was the production of large quantities of highly specific antibodies called _____ antibodies.

15. Genetic units that have been found to change their location within the chromosome are called _____.

16. Plasmids which can cause disease symptoms are called VIRULENCE plasmids.

Match the following terms with the description listed below.

Key Choices:

a.	Transformation	e.	F+
b.	Conjugation	f.	F-
c.	Specialized transduction	g.	Hfr
d.	Generalized transduction	h.	F'

1. The insertion of phage DNA into specific host chromosome sites resulting in the transduction of only certain genes.

_f_ 2. A recipient cell.

_e_ 3. A donor cell that can donate a copy of its entire chromosome.

_g_ 4. A donor cell that donates the F factor and some duplicated chromosomal genes.

_b_ 5. A genetic mechanism involving pili.

Match the following terms with the descriptions listed below.

Key Choices:

a.	Bacteriocinogens	d.	Hybridomas
b.	R - factors	e.	Gene amplification
c.	Replicons	f.	Recombinant DNA

_f_ 6. Human insulin was produced in bacteria by this process.

_b_ 7. Staphylococcal bacteria have developed resistance to antibiotics through these DNA pieces.

_d_ 8. Genetic recombinations involving cells of higher organisms.

_a_ 9. A plasmid that directs the production of a growth-inhibiting protein.

_e_ 10. The induction of plasmids to reproduce inside cells at a rapid rate.

REVIEW QUESTIONS

True/False (Mark T for True, F for False)

F 1. Transformation can only occur between a living recipient and a dead donor.

T 2. An important significance of transduction is that it suggests a mechanism for the viral origin of cancer.

T 3. Hfr's allow the transference of the largest numbers of genes from donors to recipient cells.

F 4. Hybridomas are formed by opening the cell walls of similar bacteria and allowing them to fuse.   → Fusion of eukaryotic cell.

T 5. Lysogenized bacteria contain viral prophages which can act as "timebombs."

T 6. Recombinant DNA technology as applied to industry ensures easy isolation of the product (being produced). after it had be

T 7. Mutations account for the greatest amount of genetic diversity while gene transfers contribute only a small amount of change.   some   → the greatest

Multiple Choice

b 8. Plasmids:
   a. cannot replicate on their own.
   b. are small circular pieces of DNA.
   c. represent the DNA that is incorporated into the host chromosome.
   d. are transmitted during transformation.
   e. none of the above are true.

C 9. The phage lambda that infects *Escherichia coli:*
   a. can transfer a great variety of genes.
   b. can undergo generalized transduction.
   c. can sometimes transduce specific genes. → carries
   d. always kills the host cell.
   e. all of the above are true.

C 10. An Hfr:
   a. transfers only a few genes at a time.
   b. possesses the F'pilus.
   c. transfers large numbers of genes in a linear cycle.
   d. mates only with F'cells.
   e. none of the above are true.

_d_ 11. Protoplast fusion:
  a. can only occur between strains of the same species.
  b. can only occur with prokaryotic cells.
  c. has only been attempted experimentally.
  d. can be used to obtain the best genetic characteristics of two organisms.
  e. none of the above are true.

12. Concept question:

  A research laboratory isolated and purified a new interleukin type molecule from a macrophage. This new protein seems to greatly increase the activity of cytotoxic T-cells. How could this laboratory produce enough of this molecule to study?

ANSWER KEY

Fill-in-the-Blank

1. Transformation 2. Competence factor 3. Bacteriophages
4. Temperate 5. Plasmids 6. F pilus 7. Hfr 8. R factors
9. Colicins 10. Protoplast fusion 11. Gene amplification
12. Recombinant DNA 13. Replicon 14. Monoclonal 15. Transposons
16. Virulence

Matching

1.c 2.f 3.g 4.h 5.b

6.f 7.b 8.d 9.a 10.e

Review Answers

1. **False** Transformation can occur with a living or dead donor cell. With a dead cell, the wall is usually disrupted and the DNA can be automatically released. With a live donor, small sections of DNA may be released especially during the log phase of reproduction. (pp. 196-199)

2. **True** The incorporation of prophages into bacterial chromosomes illustrates the close evolutionary relationship between the virus DNA and the host DNA. In order for incorporation to occur, there must be similar base sequences. (pp. 200-202)

3. **True** High frequency of recombination strains can allow the transfer of the entire host chromosome into the recipient whereas other mechanisms can only transfer short segments.
(pp. 203-205; Figure 8.9)

4.   **False**   Hybridomas are normally formed by the fusion of two eukaryotic cells, one a normal cell and the other a cancer cell. Bacterial protoplast fusion can occur if the cell walls are removed, protoplasts formed, and then the cells fused. (pp. 216-217)

5.   **True**   Some prophages can incorporate into the host DNA without replication and destruction of the bacterial cell. However, these phages can be induced by a number of means to reenter the lytic cycle. (pp. 200-201; Figure 8.3)

6.   **False**   A major problem encountered in industry when using recombinant DNA technology is to isolate the products after they are made. Although the product is made in large quantity it must still be purified. (p. 213)

7.   **False**   Mutations account for some genetic diversity, however, gene transfer provides the greatest increases in the genetic diversity of organisms. (p. 196)

8.   **b**   Plasmids can replicate on their own; are extrachromosomal pieces of DNA; and are transferred during conjugation. (pp. 206-208; Figure 8.11)

9.   **c**   The phage lambda carries out specialized transduction; can only transfer a few genes; must insert at a specific location; and is classified as a temperate phage. (p. 201; Figure 8.4)

10. **c**   An Hfr can transfer a large number of genes; possesses the F pilus; and mates with F-cells. (pp. 203-205; Table 8.2)

11. **d**   Protoplast fusion can occur between two different species of the same genus; can occur with prokaryotic and eukaryotic cells; and has been used to commercially produce new antibiotics. (p. 209)

12. Since the molecule is a protein, recombinant DNA technology could be one possible way. This would involve removal of DNA from the macrophage, cutting it into small segments, and incorporating the desired DNA pieces into a replicon such as a plasmid. The plasmid could then be cloned, reinserted into a bacterium, and the cells stimulated to divide and produce the desired protin in a sizeable quantity. (p. 210)

CHAPTER 9

MICROBES IN THE SCHEME OF LIFE: AN INTRODUCTION TO TAXONOMY

Taxonomy is probably not one of the most exciting topics in science. Perhaps only the most devoted science students can really glean enjoyment over this subject. Actually, years ago, it probably was even worse since all that was available were long laborious dichotomous keys based primarily on physical and biochemical characteristics. This meant having to memorize large volumes of rather boring taxonomic schemes.

Today, as this chapter will describe, taxonomy has changed quite a bit. We still have the same dichotomous keys with many of the same taxonomic units such as kingdoms, phyla, classes, and families. But the methods used in taxonomy have vastly changed! Although physical and biochemical characteristics are still used, other more accurate and revealing methods have now been incorporated. These include genetic homology, DNA hybridization, protein profiles, and even immunology. These new methods have created considerable interest and even controversy as major questions about evolutionary relationships have been answered but new questions have been posed. Linnaeus would certainly be proud if he were alive today.

STUDY OUTLINE

I. **Taxonomy - The Science of Classification**
   A. Linnaeus - the father of taxonomy
   B. Using a taxonomic key
   C. Problems in taxonomy
   D. Developments since Linnaeus
II. **The Five-Kingdom Classification System**
   A. General features
   B. Kingdom Monera
   C. Kingdom Protista
   D. Kingdom Fungi
   E. Kingdom Plantae
   F. Kingdom Animalia
III. **Classification of Viruses**
IV. **The Search for Evolutionary Relationships**
   A. Special methods needed for prokaryotes
   B. Numerical taxonomy
   C. Genetic homology
   D. Other techniques
   E. Significance of findings

REVIEW NOTES

A. How are microorganisms named?

Organisms are named according to several characteristics such as where they are found, who discovered them, and what diseases do they cause (see Table 9.1). In science however, it is essential to have accurate and standardized names. The science of classification is called taxonomy. This system insures that we name organisms based on fundamental concepts of unity and diversity among living organisms.

B. What did Linnaeus contribute to taxonomy?

Linnaeus is credited with a system of classification called binomial nomenclature which specifically identifies each living organism and provides them with a last or genus name, and a first or species name. He also established the basic hierarchy of taxonomy of families, orders, classes, phyla, and kingdoms. See Table 9.2 for examples of how this classification scheme is used.

C. How is a dichotomous taxonomic key used to identify organisms?

Taxonomic keys are used to identify organisms according to physical and/or biochemical characteristics. A dichotomous key uses a series of paired statements that are presented as either/or choices. By selecting the appropriate statement, an organism can be classified rather quickly. If the key is sufficiently detailed and if enough information is known about the organism, it can be identified down to the genus and species. See Figure 9.2 for an illustration of how a dichotomous key is used in identification.

| 1a | Grade point average 3.0 or better | Go to 2 |
| 1b | Grade point average less than 3.0 | Go to 3 |

| 2a | Study at least 20 hours per week | Hardworking good student |
| 2b | Study less than 20 hours per week | Lucky good student |

| 3a | Study at least 25 hours per week | Hardworking not-so-good student |
| 3b | Study less than 25 hours per week | More work might make you a good student! |

Figure 9.2: Dichotomous key.

D.   What are some problems and developments in taxonomy since Linnaeus?

Several problems have developed since Linnaeus. One major problem has been that evolutionary changes occur at rapid rates and our knowledge of the evolutionary history of organisms is incomplete.   Therefore any taxonomic system must be flexible to accommodate the new discoveries that are made. Another problem has been in deciding what constitutes a kingdom and what constitutes a species.

Major changes in taxonomy since Linnaeus have been in the formulation of five kingdoms rather than the two he proposed. Also, a descriptive taxonomic scheme edited by Bergey was developed to accommodate the increasing variety of bacteria. This Manual of Determinative Bacteriology has also undergone several revisions as a result of new evolutionary evidence.

E.   What are the main characteristics of the kingdoms in the five-kingdom system of taxonomy?

The five kingdoms proposed by Whittaker are Monera, Protista, Fungi, Plantae, and Animalia. See Table 9.3 for a summary of the basic characteristics of these five kingdoms.

Figure 9.4 illustrates the interrelationships of these five kingdoms.   Note that the Monera and Protista provide an evolutionary base for the other three kingdoms.

F.   How are viruses classified?

Viruses are acellular particles that consist of nucleic acids and a protein coat. They are not assigned to any kingdom and there is much debate over whether they are even "alive." Currently they are classified on the basis of their nucleic acid, their protein coat, and the presence of an envelope.

TABLE 9.3 The five-kingdom system
of classification

Kingdom	Characteristics
Monera	Prokaryotic; unicellular but sometimes cells are grouped; nutrition by absorption, but in some forms by photosynthesis or chemosynthesis; reproduction asexual, usually by fission.
Protista	Eukaryotic; unicellular but sometimes cells are grouped; nutrition varies among phyla and can be by ingestion, photosynthesis, or absorption; reproduction asexual and in some forms both sexual and asexual.
Fungi	Eukaryotic; unicellular or multicellular, nutrition by absorption; reproduction usually both sexual and asexual and often involves a complex life cycle.
Plantae	Eukaryotic; multicellular; nutrition by photosynthesis.
Animalia	Eukaryotic; multicellular; nutrition by ingestion but in some parasites by absorption; reproduction primarily sexual.

G. What special methods are needed for determining evolutionary relationships among prokaryotes?

Prokaryotic organisms have few morphological characteristics and have left only a sparse fossil record. Consequently, special methods are needed for determining their relationships. These methods include numerical taxonomy, genetic homology, and other techniques.

With numerical taxonomy, organisms are compared on a large number of characteristics such as Gram staining, morphology, types of enzymes, and oxygen requirements. See Figure 9.12 for an example of how comparisons are made. Genetic homology involves comparisons based on the cell's DNA. Various techniques have been developed to accomplish this. These include nucleotide base comparisons involving G-C ratios, DNA hybridization involving the matching of strands of DNA between two organisms, DNA and RNA sequencing, and protein profiles involving a study of the amino acid sequences of certain proteins.

Other techniques include studies of the ribosomes of organisms, immunological reactions, and phage typing.

LEARNING ACTIVITIES

Complete each of the following items by supplying the appropriate word or phrase.

1. The science of classification is known as _____.

2. The use of a two-name system for each organism is called
   Binomial _nomenclature_.

3. The "first" name of an organism that is never capitalized is the ___Species___.

4. A taxonomic key that uses paired statements describing characteristics is a _____ key.

5. The kingdom that includes all the prokaryotes is the ___Monera___.

6. Eukaryotic organisms that are all unicellular are placed in the ___Protista___.

7. Prokaryotic, blue-green algae are known as the ___Cyanobacteria___.

8. A subgroup of a species that differs in one or more characteristics from other subgroups is a ___strain___.

9. Primitive bacteria that lack peptidoglycan and are found only in very extreme environments are classed as _____.

10. A tiny virus that consists only of fragments of RNA is a ___Viroids___.

11. Huge fossilized mats of ancient prokaryotes are called _____.

12. Comparing organisms on a large number of characteristics and grouping them according to the % of shared characteristics is called _____ taxonomy.

13. Taxonomic studies based on their DNA similarities are called _____ _____.

14. A taxonomic method where studies are made of the degree of annealing between strands of DNA from test organisms is _____ _____.

15. A taxonomic method where studies of the similarities of cellular proteins are made is _____ _____.

16. A taxonomic method where bacterial strains are compared based on the type of bacteriophages that can infect them is called _____ _____.

17. The taxonomic "Bible" of the prokaryotes is _____ manual.

18. The "Father" of taxonomy is ___Linnaeus___.

19. A kingdom of organisms that obtains nutrients solely by absorption is the ___Fungi___.

20. Acellular entities that are not classified in any kingdom are
    Viruses .

21. A DNA fragment that has sequences complementary to those being
    sought could be called a DNA _____.

Match the following terms with the description listed below.   The
choices may be used more than once.

   Key Choices:

        a. Protista   f. Archaebacteria
        b. Monera   g. Cyanobacteria
        c. Fungi   h. Viruses
        d. Animals   i. Viroids
        e. Plants

___e__ 1.   Eukaryotes that can undergo photosynthesis.

___c__ 2.   Eukaryotes that obtain their nutrients solely by
            absorption.

___f__ 3.   Primitive prokaryotes found around undersea volcanic
            vents.

___g__ 4.   Photosynthetic prokaryotes.

___h__ 5.   Acellular entities classified by their nucleic acids.

___i__ 6.   Acellular entities that consist only of fragments of RNA.

Match the following terms with the descriptions listed below.

   Key Choices:

        a. Numerical taxonomy   c. DNA hybridization
        b. Genetic homology   d. Protein profiles
                                    e. DNA & RNA sequencing

_____ 7.   Involves C-G ratios to determine relatedness.

_____ 8.   Involves the PAGE technique.

_____ 9.   Organisms are grouped on the basis of Gram reaction,
            morphology, and numerous biochemical tests.

___c__ 10. Involves matching DNA strands by separating and
            reannealing them.

REVIEW QUESTIONS

True/False (Mark T for True, F for False)

_____ 1.   The animal kingdom is the only kingdom that lacks any organisms that relate to microbiology.

_____ 2.   Using proper taxonomy, the genus and species names are always italicized and capitalized.

_____ 3.   A dichotomous key always requires paired statements.

_____ 4.   Viruses are currently classified according to their nucleic acids, their capsid, and the presence or absence of envelopes.

__T__ 5.   Quite often, the genus name of an organism is used to honor the scientist credited with its discovery.

_____ 6.   In DNA hybridization, a high degree of similarity exists when the DNA of two organisms have long identical sequences of bases.

_____ 7.   The ATCC represents a branch of the CDC whose task is to identify new diseases.

Multiple Choice

__C__ 8.   Viruses are:
   a.   considered to be prokaryotes.
   b.   not classified by any classification scheme.
   c.   considered to be acellular particles.
   d.   placed in the kingdom Monera.
   e.   none of the above are true.

__a__ 9.   The scientist credited with the development of the binomial scheme is:
   a.   Linnaeus   c.   Leeuwenhoek
   b.   Semmelweiz   d.   Aristotle

__C__ 10.   The scientist most responsible for establishing the five-kingdom classification system currently in use is:
   a.   Linnaeus   c.   Whittaker
   b.   Bergey   d.   Margulis

__d__ 11.   A characteristic associated only with the kingdom Prokaryotae:
   a.   cell membrane   c.   photosynthesis
   b.   unicellularity   d.   lack a true nucleus

85

___b___ 12. Which of the following kingdoms are characterized by members that are eukaryotic and mostly all unicellular.
   a. Monera   c. Fungi
   b. Protista   d. Plantae
                                e. Animalia

ANSWER KEY

Fill-in-the-Blank

1. Taxonomy 2. Binomial nomenclature 3. Species
4. Dichotomous key 5. Monera 6. Protista 7. Cyanobacteria
8. Strain 9. Archaebacteria 10. Viroid 11. Stromatolites
12. Numerical taxonomy 13. Genetic homology 14. DNA hybridization
15. Protein profiles 16. Phage typing 17. Bergeys 18. Linnaeus
19. Fungi 20. Viruses 21. Probe

Matching

1.e 2.c 3.f 4.g 5.h/i 6.i

7.b 8.d 9.a 10.c

Review Answers

1.   **False**   Several groups of animals are of interest to micro-biologists since some act as parasites such as the tapeworms and roundworms which live on or in other animals and some act as carriers of microbial disease agents such as certain insects. (p. 231; Figure 9.9)

2.   **False**   The genus name is always capitalized and underlined or italicized and the species name is only underlined or italicized. (p. 224)

3.   **True**   A dichotomous key always uses paired statements to present an "either or" choice such that only one statement can be true. (pp. 225-226; Figure 9.3)

4.   **True**   Viruses, since they are acellular particles, require a separate classification system.   Currently the scheme uses the three parameters mentioned. (p. 232; Figure 9.10)

5.   **True**   Many scientists including Pasteur, Rickets, Lister, Yersin, Neisser, and Escherich all are honored in this manner. (p. 225; Table 9.1)

6.   **True**   When the DNA of two organisms are split apart and then allowed to reanneal, the amount of relatedness can be determined. The greater the amount of base pairing by the strands the more they are related. (p. 235; Figure 9.14)

7.   **False** ATCC refers to American Type Culture Collection which is based in Rockville, Maryland. The company maintains thousands of lyophilized cultures to ensure their exact characteristics. (p. 225)

8.   c   Viruses are acellular particles and are classified on the basis of their nucleic acid, capsid symmetry, and presence of an envelope. They are not placed in any of the five kingdoms. (p. 232)

9.   a   Semmelweiz helped introduce asepsis in obstetrics, Leeuwenhoek was the first to observe microbes, and Aristotle was a Greek philosopher who formulated important concepts of science. (p. 224; Figure 9.1)

10.   c   Linnaeus developed the binomial system; Bergey edited the first work on taxonomy of bacteria; Margulis proposed a four-kingdom system with protoctista rather than Protista. (p. 227)

11.   d   All living cells possess cell membranes; most members of the kingdom Protista are unicellular; members of the Plant kingdom can photosynthesize. (p. 229; Table 9.3)

12.   b   Members of the kingdom Monera are all prokaryotic; members of the kingdom Fungi are eukaryotic but mostly multi-cellular; members of the kingdom Plantae are eukaryotic but mostly multicellular; and members of the kingdom Animalia ar eukaryotic but mostly multicellular. (pp. 228-231; Table 9.3)

CHAPTER 10

THE BACTERIA

Most beginning students of microbiology have little appreciation for the incredibly rich diversity of microbes found on this earth. Even those with extensive backgrounds in the field are often amazed that new organisms are continually being discovered as cultivation and isolation techniques are improved and as new environments are being discovered.

With this in mind, this chapter provides an excellent survey of the important groups of microbes that will be discussed in later chapters on disease. Therefore, it will be helpful to have an initial idea of how they are classified, what basic characteristics they have, and what diseases they cause. In order to get a sense of the most recent taxonomic schemes presented in Bergey's manual, the bacteria will be presented and described by sections. The last section provides an interesting survey of the environmentally important organisms that have unusual characteristics or have made a significant impact on our lives. This information although not essential for medical knowledge will help relate microbiology to many other fields of science.

If possible, take a moment to read over the essay at the end of this chapter. It asks the question, "are we still discovering new organisms"? This world of ours still contains hundreds of undiscovered and untested environments which could reveal microbes with characteristics never before seen. Clearly, Bergey's Manual will probably never be "finished".

STUDY OUTLINE

REVIEW NOTES

A. What criteria are used for classifying bacteria?

Because many bacteria have similar shapes and sizes, several criteria are necessary for classification of the bacteria. These include cell morphology, staining properties (especially Gram stain), growth, nutrition, physiology, biochemistry, and genetics. See **Table 10.1** for a description of these criteria. These criteria can be used to classify bacteria into a genus and, in some cases, into species and even strains.

B. What problems are associated with bacterial taxonomy?

Ever since the first manual of determinative bacteriology was written, microbiologists have been unable to agree on exactly how the members of the kingdom Monera (Prokaryotae) should be divided. Because of recent advances in taxonomic technology, many changes have been made resulting in an extremely wide diversity of opinions. Those that are looking from the top down, that is, from the kingdom level down can assign the bacteria to several divisions but have difficulty with classes and orders. Those that are looking from the bottom up, that is, from the genus level up have established individual groups or sections but have difficulty with many of the organisms within the sections. Recent data have created even more discrepancies in assignments. There is even more trouble when considering orders and classes. Hopefully, with additional information that should be obtained soon, some of the questions will be answered.

TABLE 10.1 Criteria for classifying bacteria

Morphology	Size and shape of cells, arrangement in pairs, clusters or filaments, presence of flagella, pili, endospores, capsules
Staining	Gram-positive, gram-negative, acid-fast
Growth	Characteristics in liquid and solid cultures, colony morphology, development of pigment
Nutrition	Autotrophic, heterotrophic, fermentative with different products, energy sources, carbon sources, nitrogen sources, needs for special nutrients
Physiology	Temperature (optimum and range), pH (optimum and range), oxygen requirements, salt requirements, osmotic tolerance, antibiotic sensitivities and resistances
Biochemistry	Nature of cellular components such as cell wall, RNA molecules, ribosomes, storage inclusions, pigments, antigens
Genetics	Percentages of DNA bases, DNA hybridization

C.   What is the history and significance of Bergey's Manual?

Bergey's Manual of Determinative Bacteriology is the taxonomic "Bible" for microbiology. It was first published in 1923 with David H. Bergey as the chairperson of the editorial board. It has been revised numerous times, the ninth edition being the latest. The manual contains the names and descriptions of organisms and diagnostic keys and tables for identifying each organism (see Figure 10.2).

D.   What are the characteristics of those bacterial genera that have medical significance and what diseases do they cause?

See Table 10.5 for a complete summary of the characteristics of all of the medically important members of the sections of bacteria as defined in the latest edition of Bergey's Manual. Each of these organisms will be discussed in detail in later chapters. It will be very helpful at this time to prepare a brief list of the most important members. Use Table 10.5 to set up your study list or use the study outline as provided in the beginning of this chapter review.

Also, see **Table 10.3 below** for a useful comparison of bacteria, rickettsia, chlamydia, mycoplasma, ureaplasma, and viruses.

TABLE 10.3 Characteristics of typical bacteria, rickettsias, chlamydias, *Mycoplasma*, *Ureaplasma*, and viruses

Characteristic	Typical Bacteria	Rickettsias	Chlamydias	*Mycoplasma*	*Ureaplasma*	Viruses
Cell wall	Yes	Yes	Yes	No	Sometimes	No
Grow only in cells	No	Yes	Yes	No	No	Yes
Require sterols	No	No	No	Sometimes	Yes	No
Contain DNA and RNA	Yes	Yes	Yes	Yes	Yes	No
Have metabolic systems	Yes	Yes	Yes	Yes	Yes	No

Once you complete this chapter, you should be able to list the 16 medically important genera, list their basic characteristics, and list the important genera and the diseases they cause.

LEARNING ACTIVITIES

Complete each of the following items by supplying the appropriate word or phrase.

1. A strain of bacteria that represents the standard of the species is the _____.

2. According to the current taxonomy, bacteria have been classified into 30 groups called _____.

3. The taxonomic "Bible" used in classifying all the bacteria is _____ _____.

4. Helical shaped, motile bacteria may be called *Spirochetes*

5. Cork-screw shaped spirochetes that include the organisms that cause syphilis are called _____.

6. Aerobic motile rods that often produce diffusible yellow-green pigments and have polar flagella are_____.

7. Bacteria that inhabit the intestinal tract of humans are often referred to as ___Enteric___ ___Bacteria___.

8. Organisms that spontaneously lose their walls and form irregularly shaped cells are _____.

9. Organisms that are short curved, comma-shaped rods are often called ___vibrios___.

10. Two groups of obligate intracellular parasites that are classified in section 9 are the _____ and _____.

11. The chlamydia often form aggregations of large metabolically active structures within cells called _____ _____.

12. Small dense infectious bodies formed by the chlamydia are called _____ bodies.

13. Mycoplasmas, when grown on agar, form characteristic _____ _____ appearing colonies.

14. Bacteria that can completely hemolyze blood cells exhibit ___Beta___ hemolysis.

15. Streptococci that cause infections of the valves and lining of the heart and are alpha hemolytic are classed in the _____ group.

16. Regular, nonsporing Gram-positive rods that are used in the production of yogurt and sourdough ___lactobaci___.

17. Irregular, nonsporing Gram-positive rods that are usually club-shaped are often called _____.

18. Filamentous members of the irregular, nonsporing Gram-positive rods are the ___Actinomyces___.

19. Gram-positive, nonmotile, pleomorphic, aerobic, filamentous rods which are related to the actinomycetes and mycobacteria are _____.

20. A genus that contains soil-dwelling organisms that resemble fungi and produce a great variety of antibiotics is _____.

21. A term used to identify all members of the Enterobacteriaceae could be ___coliforms___.

Match the following terms with the description listed below.

Key Choices:

a.	Section 1	f.	Section 12
b.	Section 4	g.	Section 13
c.	Section 5	h.	Section 15
d.	Section 6	i.	Section 16
e.	Section 9	j.	Section 29

_____ 1.   Members include the staphylococci and streptococci.

_____ 2.   Members have produced 100's of different antibiotics.

_____ 3.   Includes the agents of syphilis and Lyme disease.

_____ 4.   Includes agents that infect burn wounds, and cause gonorrhea, and whooping cough.

_____ 5.   Includes the most common urinary tract invaders and agents of human plague and typhoid fever.

_____ 6.   Agents cause leprosy and tuberculosis.

_____ 7.   Includes the genera, *Bacteroides* and *Fusobacterium* that cause oral and wound infections.

_____ 8.   Members cause tetanus, anthrax, and gas gangrene.

_____ 9.   Are commonly club-shaped, pleomorphic, and filamentous.

_____ 10. Members cause Rocky Mountain Spotted Fever and typhus.

REVIEW QUESTIONS

True/False (Mark T for True, F for False)

_____ 1.   The current edition of Bergey's manual separates the bacteria into taxonomic units called parts.

_____ 2.   Bacteria are commonly grouped into phyla, classes, and orders.

_____ 3.   Orders always end in -ales.

_____ 4.   A major difference between chlamydias and mycoplasmas is that the mycoplasmas can grow on artificial culture media.

_____ 5.   Section five probably contains the greatest variety of medically important microbes including *Escherichia*, *Shigella*, and *Salmonella*.

Multiple Choice

_____ 6.   A major problem in establishing clearly defined taxonomic classes is:
a.   the bacteria are too complex.
b.   some bacteria have true nuclei.
c.   too little is known about the evolutionary relationships.
d.   the bacteria mutate too often.

_____ 7.   Methods used to classify bacteria include all the following except:
a.   staining properties.
b.   percentages of DNA bases.
c.   number of mitochondria.
d.   type of nutrition.
e.   All of the above are used in classification.

_____ 8.   Rickettsias exhibit all the following except:
a.   require sterols in their cell walls.
b.   grow only inside cells.
c.   possess complex metabolic systems.
d.   contain DNA and RNA.

_____ 9.   Section 13, the endospore forming Gram-positive rods and cocci includes the genera:
a.   *Staphylococcus* and *Streptococcus*
b.   *Mycobacterium*
c.   *Pseudomonas* and *Brucella*
d.   *Bacillus* and *Clostridium*

_____ 10. The Gram-negative aerobic rods and cocci:
a.   include the genera *Treponema* and *Borrelia*.
b.   are in Section 16.
c.   form long, acid-fast filaments and cause severe respiratory infections.
d.   are obligate intracellular parasites.
e.   None of the above are true.

94

ANSWER KEY

Fill-in-the-Blank

1. Type strain 2. Sections 3. Bergey's manual 4. Spirochetes
5. Treponemes 6. Pseudomonads 7. Enteric bacteria 8. L-forms
9. Vibrios 10. Rickettsia, chlamydia 11. Inclusion bodies
12. Elementary bodies 13. Fried egg 14. Beta hemolysis
15. Viridans group 16. Lactobacilli 17. Corynebacteria
18. Actinomyces 19. Nocardioforms 20. Streptomyces 21. Coliforms

Matching

1.f 2.j 3.a 4.b 5.c 6.i 7.d 8.g 9.h 10.e

Review Answers

1.   **False**   A previous edition had established taxonomic units called parts.   The current edition uses sections. (p. 246)

2.   **False**   Taxonomic units used in classifying bacteria include genera, species, families, classes, orders but no phyla. (p. 244)

3.   **True**   Orders end in -ales and families end in -aceae. (p. 246, appendix)

4.   **True**   Although mycoplasmas lack cell walls they still can grow on artificial media. (pp. 253-254; Table 10.3)

5.   **True**   Section 5 represents the facultative anaerobic Gram-negative rods and includes a wide variety of related genera. (p. 258; Table 10.5)

6.   **c**   Although bacteria are complex, their complexity is not necessarily an impeding factor to classification, often it is an asset; all bacteria lack true nuclei; mutational changes do not create a significant taxonomic problem. (p. 244)

7.   **c**   Since bacteria lack any complex organelles such as mitochondria, this parameter is never used in classification. (pp. 244-245; Table 10.1 and 10.2)

8.   **a**   Rickettsias and chlamydias do not require sterols in their cell walls, but many mycoplasmas and all ureaplasmas do require them. (pp. 252-253)

9.   **d**   The genera *Bacillus* and *Clostridium* contain important endospore-forming members including ones that cause anthrax, tetanus, and gas gangrene. (pp. 255-256)

10.   **e**   These members are in Section 4 and include the genera, *Pseudomonas*, *Neisseria*, *Brucella*, and *Bordetella*. (pp. 248-249)

CHAPTER 11

VIRUSES

What are viruses?   A question no doubt posed by students, teachers, scientists, and certainly by those who have contracted a serious viral disease such as rabies, polio, or AIDS.   As we begin to learn more about these particles we seem to uncover even more questions.   As we conquer one viral disease like smallpox, another seems to take its place.   As soon as the basic structures, methods of replication, and methods of cultivation of viruses are understood, new and even more strange particles appear such as the viroids and prions.   With the discovery of oncogenes in living cells, the imagination is once again stretched to the limit in terms of their possible impact on all of us.   The field of virology is once again an exciting and challenging science.

The first part of this chapter provides an overview of the most common viral characteristics and their method of classification.   This is followed by a presentation of the recently discovered particles, the prions and viroids.   The last part of the chapter is devoted to the methods of viral replication and cultivation with a brief discussion of the role and importance of viruses as teratogenic or embryonic defect causing agents.

I.   General Characteristics of Viruses
      A.   What are viruses?
      B.   Components of viruses
      C.   Shapes and sizes
      D.   Host ranges and specificity of viruses
II.   Classification of Viruses
      A.   Basic methods
      B.   RNA viruses
      C.   DNA viruses
III. Viroids and Prions
      A.   Viroids
      B.   Prions
IV.   Viral Replication
      A.   General characteristics of replication
      B.   Replication of bacteriophages
      C.   Lysogeny
      D.   Replication of animal viruses
V.   Culturing of Animal Viruses
      A.   Development of culturing methods
      b.   Types of cell cultures
VI.   Viruses and Teratogenesis
      A.   Teratogenesis
      B.   Viral groups responsible

REVIEW NOTES

A.   What are the general properties of viruses?

Viruses are acellular, infectious particles that are too small to be seen with a light microscope. They consist of a nucleic acid either DNA or RNA, but not both, and a protein coat. Some viruses also have an envelope that they acquired from cells they previously infected. Viruses range in size from 5 to 300 nm and have an icosahedral, helical, or complex shape. **See Figure 11.2** which illustrates some of the variations in shapes and sizes of viruses.

Viruses also have a very specific host range in that they generally only infect one kind of organism. They also have a specificity in that most can infect only one kind of cell.

B.   How are viruses classified?

Viruses are classified on the basis of their type of nucleic acid, their structure, presence of an envelope, chemical and physical characteristics, method of replication, and host range. See Table 11.1 for a description of how they are classified. Appendix C summarizes the current listing of families as identified by the International Committee on Taxonomy of Viruses (ICTV).

RNA viruses are divided into five classes, and DNA viruses are divided into three classes. Virtually all DNA and RNA classes contain medically important human viruses.

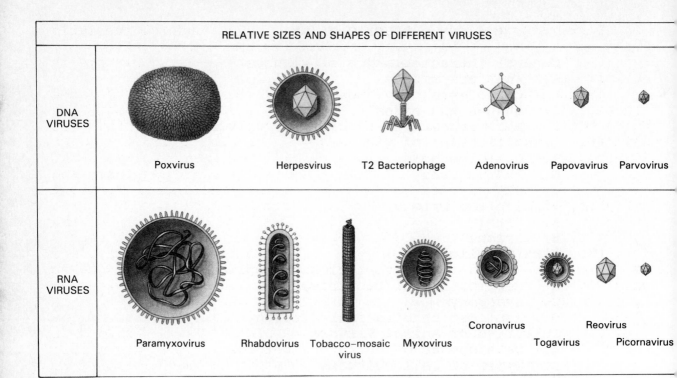

RELATIVE SIZES AND SHAPES OF DIFFERENT VIRUSES					
DNA VIRUSES	Poxvirus	Herpesvirus	T2 Bacteriophage	Adenovirus	Papovavirus · Parvovirus
RNA VIRUSES	Paramyxovirus	Rhabdovirus · Tobacco-mosaic virus	Myxovirus	Coronavirus · Togavirus	Reovirus · Picornavirus

Figure 11.2: Variations in shapes and sizes of viruses.

C.   What are the properties of viroids and prions?

Viroids are infectious particles smaller than viruses that lack a protein coat, consist of RNA, and are self-replicating. They appear to infect only plant tissues but are not well understood.

Prions are proteinaceous infectious particles believed to cause certain slow viral diseases such as Creutzfeldt-Jacob disease and kuru. Their exact origin, role, method of replication, and disease-causing potential is not clearly known.

D.   How do viruses replicate in general?

Viral replication involves five basic steps: (1) adsorption or attachment to the host cell; (2) penetration, which refers to the entry of the virus into the cell; (3) synthesis, which refers to the making of the nucleic acids and protein components; (4) maturation, which refers to the assembly of the components; and (5) release, which is the departure of the intact virus from the cell.

E.   How do lytic and temperate bacteriophages replicate?

Bacteriophages can either lyse the cell and undergo the lytic cycle or they can reside within the cell and establish a lysogenic cycle. See Figure 11.9 for an illustration of these two cycles noting the interrelationship.

Lytic phages, which include the T-even phages, follow the classic five steps of replication. They possess recognition factors that allow them to attach or adsorb to a specific cell surface. Penetration is accomplished by means of an enzyme to weaken the wall and a syringe tube-like structure to inject the nucleic acid. The virus then takes over the cell's machinery and directs the construction or synthesis of all the necessary viral parts. Maturation occurs with the assembly of all the viral components in a very orderly fashion. With the completion of the intact viruses, an enzyme is produced and lysis of the cell occurs followed by release of the phages. Figure 11.10 illustrates the growth curve for a bacteriophage. Note that a complete cycle takes only about 30 minutes with the release of hundred's of viruses! Lytic bacteriophage viruses grown on a bacterial lawn will form clear areas called plaques which are filled with viruses.

Lysogenic or temperate phages do not immediately kill the cell but generally establish a long-term relationship. They can either exist as a prophage or revert to a lytic phage in the future. This relationship is important because the virus can provide extra genes for the bacterial cell such as the ability to produce specific toxins or enzymes.

F. How do animal viruses replicate?

Animal viruses follow the same basic steps found with bacteriophages but generally have a greater variety of mechanisms. Adsorption involves receptors on the viral surface which allows attachment to the cell membrane. Penetration follows by a process such as pinocytosis. Synthesis and maturation differ in DNA and RNA viruses. See Figure 11.14 for a description of the process in DNA viruses and Figure 11.15 for the process in RNA viruses.

DNA viruses have a traditional mechanism with the DNA being transcribed into mRNA which then directs the protein synthetic mechanisms. RNA viruses exhibit a great variety of mechanisms depending on whether the virus is single stranded or double stranded and negative or positive sense RNA. In some cases, the RNA virus requires a reverse transcriptase to form DNA, which is then used to direct the synthesis of components. In some cases, the viral DNA is incorporated into the cell's DNA.

Viruses are assembled in the cell and released by direct lysis or by budding through the membrane.

G. How were methods developed to culture animal viruses?

The development of cultural methods proceeded very slowly because of difficulty in developing tissue cultures and keeping them free of bacterial contamination. Antibiotics and special enzymes aided in the development of useful culture systems.

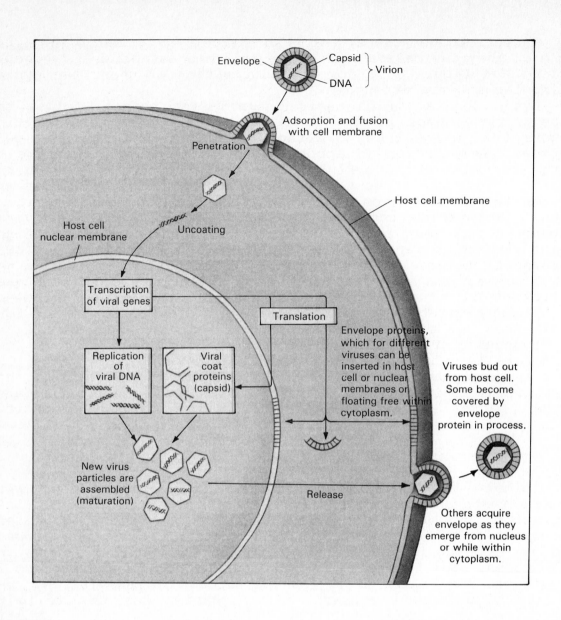

Figure 11.14: Replication of a DNA animal virus.

H.   What types of viral cultures are currently in use?

Several cell culture systems are in use today and include

primary cell cultures, diploid fibroblast strains, and continuous cell lines. The primary cell cultures come directly from animals, can not be subcultured, and normally do not last long. They can be used to support the growth of a wide variety of viruses. Diploid fibroblast strains are the most widely used because they can be maintained for several years. They are often used in making vaccines. Continuous cell lines are usually derived from cancer cells and can grow almost indefinitely. They are commonly used in research.

I.   **What is a teratogen and how do viruses act as teratogens?**

Teratogens are agents that can induce embryonic defects. Viruses can act as teratogens by crossing the placenta and infecting embryonic cells. The greatest damage seems to occur in the early developmental stages. Virtually any body organ or tissue can be affected. Rubellaviruses, cytomegaloviruses, and herpesviruses seem to be the most common viral agents.

LEARNING ACTIVITIES

Complete each of the following items by supplying the appropriate word or phrase.

1.   Since viruses can only grow inside living cells they are called ___obligate___ ___intracellular___ parasites.

2.   The viral nucleic acid is covered by a protein coat called a ___capsid___.

3.   Some viruses have a lipid bilayer around them called an ___envelope___.

4.   Viruses that exhibit a many-sided box-like capsid are called ___icosahedral___ viruses.

5.   Viruses are usually measured in metric units called _____.

6.   The different kinds of organisms a virus can infect is called its ___hosts___ ___Range___.

7.   A nucleic acid strand that encodes the information for making proteins needed by a virus is called _____ _____ nucleic acid.

8.   Viruses are primarily classified on the basis of their ___nucleic___ ___Acid___.

9.   An enzyme that certain RNA viruses possess to copy RNA into DNA is a _____ _____.

10. Viral-like agents that may only contain infectious protein particles are ___Prions___.

11. Viruses that lack capsids, contain RNA, and infect plant tissue may be ___Viroids___.

12. The first step in the viral replication process is ___adsorption___.

13. Virus particles are assembled inside the host cell during the ___maturation___ phase.

14. The time from adsorption to release of phage particles is the ___burst___ time.

15. During the ___Eclipse___ phase, phages disappear from the culture due to absorption and/or penetration into the host cell.

16. Clear areas on a bacterial lawn caused by lytic viruses are ___Plaques___.

17. The stimulation of a temperate phage to become virulent is called ___Induction___.

18. _____ is the process whereby animal viruses are taken into a cell in a manner similar to phagocytosis.

19. Enveloped viruses obtain their coating from the _____ _____ of the host cell.

20. Viral cell cultures that come directly from the animal and are not subcultured are called _____ _____ _____.

21. When viruses cause a visible effect on their host cells, the effect is called ___cytopathic___.

22. An agent that can induce defects during embryonic development is known as a ___Teratogen___.

23. The highest catagory currently used in classification of viruses by the ICTV is ___Family___.

24. A rash illness of children caused by a parvovirus is called the _____ disease.

25. The "T" used to identify the different T-even bacteriophages (T2, T4, etc) stands for ___types___.

Match the following terms with the description listed below. Terms may be used more than once.

Key Choices:

a. Togaviruses   f. Rhabdoviruses
b. Picornaviruses   g. Orthomyxoviruses
c. Bunyaviruses   h. Reoviruses
d. Arenaviruses   i. Retroviruses
e. Paramyxoviruses

_____ 1. Enveloped, helical, negative sense, ssRNA.

_____ 2. Naked, polyhedral, positive sense, ssRNA.

_____ 3. Arthropod borne, enveloped, polyhedral, example: rubella.

_____ 4. Reverse transcriptases, example: AIDS.

_____ 5. Enveloped, helical, ssRNA, example: influenza.

Match the following terms with the description listed below. Terms may be used more than once.

Key Choices:

a. Adenoviruses   d. Papovaviruses
b. Herpesviruses   e. Parvoviruses
c. Poxviruses

_____ 6. dsDNA, largest viruses of all viral groups.

_____ 7. dsDNA, polyhedral, examples: shingles, mononucleosis, chickenpox.

_____ 8. ds, circular DNA, example: warts.

_____ 9. Enveloped viruses.

Match the following terms with the description listed below.

Key Choices:

a. HeLa   e. Lysogen
b. TORCH   f. Induction
c. Syncytia   g. Pentons
d. Plaques

_b_ 10. A series of serological tests designed to detect diseases in pregnant women and newborn infants.

_a_ 11. A continuous cell line used to grow viruses.

___C___ 12. Giant cells formed as a result of certain viral infections.

___g___ 13. Units of the capsomere.

REVIEW QUESTIONS

True/False (Mark T for True, F for False)

___F___ 1. Some of the complex viruses have both DNA and RNA in their genome.

___T___ 2. Following the release of viral particles from the cell, the cell usually, but not always, dies.

___T___ 3. The diseases of botulism and diphtheria are actually viral related since a temperate phage provides the bacterium with the ability to make a toxin.

___F___ 4. The most common childhood viral diseases are those found in the arenavirus group.

___T___ 5. A tiny viral group that includes the viruses that cause polio is the picornaviruses.

Multiple Choice

_____ 6. Viral groups known to have members that cause tumors include all the following except:
a. herpesviruses   c. papovaviruses
b. reoviruses   d. adenoviruses

_____ 7. Viruses with single-strand, negative-sense RNA:
a. cannot replicate inside cells.
b. must replicate inside the nucleus of cells.
c. must first make a positive sense RNA strand
d. do not require a transcriptase to make viral components.

_____ 8. Viroids differ from viruses in all the following ways except:
a. viroids lack capsids.
b. viroids possess RNA and infectious protein units.
c. viroids possess nucleic acids of low molecular weight.
d. viroids are hard to identify in tissue.

_____ 9.   Viral groups that cause childhood infections include all
           the following except:
           a. picornaviruses   d.   reoviruses
           b. herpesviruses   e.   paramyxoviruses
           c. rhabdoviruses

__d__ 10. Members of the herpesviridae include all the following
           except:
           a. varicella   c.   cytomegalloviruses
           b. zoster   d.   rubellaviruses

__c__ 11. In the penetration step during replication of
           bacteriophages:
           a.   tail fibers contract to push the internal tube
                through the wall.
           b.   the entire phage unit penetrates through the cell
                wall.
           c.   lysozyme in the phage tail weakens the bacterial cell
                wall.
           d.   cell wall receptor sites act to bind the tail core
                proteins which osmotically opens the wall.

ANSWER KEY

Fill-in-the-Blank

1. Obligate intracellular 2. Capsid 3. Envelope 4. Icosahedral
5. Nanometers 6. Host range 7. Positive sense nucleic acid
8. Nucleic acid 9. Reverse transcriptase 10. Prions 11. Viroids
12. Adsorption 13. Maturation 14. Burst time 15. Eclipse
16. Plaques 17. Induction 18. Viropexis 19. Cell membrane
20. Primary cell cultures 21. Cytopathic 22. Teratogen
23. Family 24. Fifth 25. Type

Matching

1.e,f 2.b 3.a 4.i 5.g

6.c 7.b 8.d 9.b,c

10.b 11.a 12.c 13.g

Review Answers

1.   **False** All viruses possess either DNA or RNA but not both.
(p. 256; Table 11.1)

2.   **True** Viruses that are released by budding through the cell
membrane may or may not kill the cell. Viruses that are released
by cell lysis will kill the cell. (p. 274)

3.   **True**   Certain bacteria require the presence of temperate phages with the genetic information that codes for the production of a toxin.   Without the phages the organisms do not cause diseases. (pp. 268-269)

4.   **False**   The arenavirus group contains members that cause hemorrhagic fevers. The picornaviruses and paramyxoviruses contain members that cause the common cold, measles and mumps. (pp. 260-261; Table 11.1)

5.   **True**   The picornaviridae are small (27-30 nm) icosahedral viruses that include the enteroviruses (polio and echo) and rhinoviruses (common cold). (p. 260; Table 11.1)

6.   b   The reoviruses generally cause respiratory and gastrointestinal infections. (p. 261; Table 11.1)

7.   c   These viruses require a transcriptase enzyme to make the necessary positive-sense nucleic acid for proper decoding by the ribosome. They also can replicate in either the cytoplasm or nucleus. (pp. 259 and 272)

8.   b   Viroids which appear to infect plants do not possess any capsid or other type of protein material. (p. 264)

9.   c   Members of the rhabdoviridae cause rabies in animals and also infect insects and fish. (p. 261; Table 11.1)

10. d   The rubellaviruses cause German measles and are members of the togaviridae which also include the encephalitis and tropical fever viruses. (pp. 263 and 260; Table 11.1)

11. c   Tail fiber contraction occurs during adsorption; only the nucleic acid penetrates the cell wall; binding to receptor sites occurs during absorption and entrance into the cell does not occur by osmosis. (p. 277)

CHAPTER 12

EUKARYOTIC MICROORGANISMS AND PARASITES

Microbiology is not just a study of organisms that cannot be seen with the naked eye. It is a field that includes such macroscopic organisms as protists, fungi, helminths, and even arthropods. Many of these representatives are certainly visible without a microscope but are considerably important as human pathogens, parasites, vectors, or reservoirs of disease. Therefore it is very important to include these microbes because, all too often, students of allied health programs will have little opportunity to explore them in any other course in their college career.

With this in mind, the chapter begins with a useful discussion of the principles of parasitology followed by a survey of the protists, fungi, helminths, and arthropods. While the medical significance of these groups is stressed, sufficient information is provided to ensure that the role of these organisms in the natural environment can be appreciated.

STUDY OUTLINE

I.   **Principles of Parasitology**
    A.   General characteristics
    B.   Significance of parasitism
    C.   Parasites in relation to their hosts

II.   **Protists**
    A.   Characteristics of protists
    B.   Importance of protists
    C.   Classification of protists

III.   **Fungi**
    A.   Characteristics of fungi
    B.   Importance of fungi
    C.   Classification of fungi

IV.   **Helminths**
    A.   Characteristics of helminths
    B.   Classification of helminths

V.   **Arthropods**
    A.   Characteristics of arthropods
    B.   Classification of arthropods

REVIEW NOTES

A.   What is a parasite, and what are the principles of parasitism?

A parasite is an organism that lives at the expense of another organism called the host. The study of parasites, which include members of the protozoa, helminths, and arthropods, is parasitology. Parasites are responsible for a major portion of human disease and death and play an important role in the worldwide human economy.

They appear to live on or in their hosts and many are obligate parasites having to spend some of their life cycle with the host. Some parasites are permanent residents others only temporary or accidental. They can reproduce sexually in their definitive hosts and spend other life stages in intermediate hosts. They can also be transmitted from a reservoir host via a vector to a human host. Most importantly, a good parasite is one that has become well adapted to its host and causes little damage.

B.   What are protists, and why are they important?

The protists are members of the kingdom Protista, are eukaryotic, microscopic, and mostly unicellular. They can be self-supporting or autotrophic or heterotrophic where they feed on other organisms. Some are parasitic.

Many are important in the food chains as producers or decomposers and many are very important economically. Some are responsible for forming calcified deposits used in building materials while others multiply so rapidly that they form thick impenetrable layers in lakes, causing many useful organisms such as fish to die.

C.   How do groups of protists differ?

Protists may be grouped in several ways. The simplest way is to group them according to the kingdom of macroscopic organisms they most resemble. **See Table 12.1** for a summary of their characteristics and examples.

TABLE 12.1 Properties of protists

Group	Characteristics	Examples
Plantlike protists	Have chloroplasts, live in moist, sunny environments	Euglenoids, diatoms, dinoflagellates
Funguslike protists	Most are saprophytes, may be unicellular or multicellular	True slime molds, cellular slime molds
Animallike protists	Heterotrophs, most are unicellular, most free-living, but some are commensals or parasites	Mastigophorans, sarcodinas, sporozoans, and ciliates

D.   What are the fungi, and why are they important?

The fungi are members of the kingdom Fungi, are eukaryotic, saprophytic (live off of dead organic material), and mostly multicellular except for the yeasts which are unicellular. They form threads called hyphae that assemble into a mat called a mycelium. Molds typically form an unorganized mycelium while mushrooms have a very organized mycelial structure. Most reproduce both sexually and asexually with their sexual stage being used for classification.

Fungi are very important as decomposers in the environment but also as parasites on plants, animals, and humans. Some fungi are economically important as foods and as producers of enzymes and antibiotics.

E.   How do groups of fungi differ?

Fungi are usually classified according to the nature of their sexual stage in their life cycle. Unfortunately, some fungi do not exhibit a sexual stage and others are difficult to match their sexual and asexual stages properly. Therefore, they will be grouped into the water molds (Oomycota), bread molds (Zygomycota), sac fungi (Ascomycota), club fungi (Basidiomycota), and imperfect fungi (Deuteromycota). See Table 12.2 for a description and examples of these groups of fungi.

F.   What are parasitic helminths, and why are they important?

The helminths, or worms, are members of the Animal kingdom, are eukaryotic, multicellular, bilaterally symmetrical, have head and tail ends, and are differentiated into tissue layers. Most of these worms are harmless, free-living, environmental or aquatic organisms. A few have established a parasitic relationship with man, animals, and plants. **See Figure 12.18 for an example.**

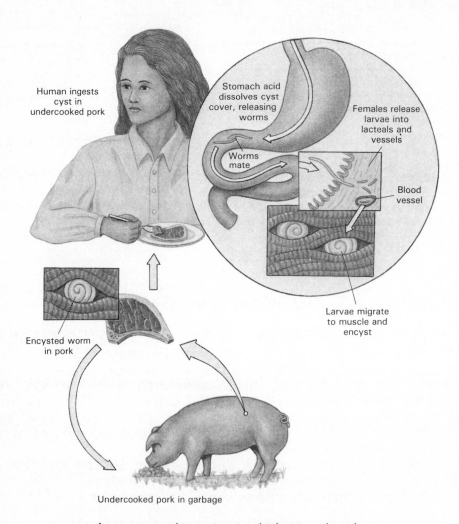

Figure 12.18: Life cycle of <u>Trichinella spirallis.</u>

G.   How do groups of parasitic helminths differ?

The helminths are commonly grouped into the flatworms and roundworms or nematodes. The flatworms include the tapeworms which look much like a flattened measuring tape and the flukes which look like a flat leaf. They all lack a coelom, have a simple digestive tract with only one opening, and are mostly hermaphroditic, that is, both sexes within the same worm. The worms commonly possess hooks and suckers at the head end which are used for attachment.

The nematodes have a pseudocoelom, separate sexes, and a cylindrical body with a tough outer covering or cuticle. They include hookworms, pinworms, and other parasitic worms of the intestinal tract, tissues, and lymphatics. See Table 12.3 for a summary of the properties of helminths.

H. What are the characteristics of parasitic and vector arthropods?

The arthropods constitute the largest group of living organisms, are members of the Animal kingdom, have jointed, chitinous exoskeletons, segmented bodies, and jointed appendages. They include the arachnids, the insects, and the crustacea. Most are free-living in the soil, on vegetation, and in fresh and salt water but some also serve as intermediate hosts for other human parasites or act directly as the parasite.

I. How do groups of parasitic and vector arthropods differ?

Parasitic and vector arthropods include some arachnids such as the ticks and mites, several insects such as flies, fleas, and mosquitoes, and a few crustacea such as crabs and copepods. See Table 12.4 for a summary of the arthropods.

Table 12.5 is an excellent summary of all the diseases transmitted by arthropods.

LEARNING ACTIVITIES

Complete each of the following items by supplying the appropriate word or phrase.

1. An organism that lives at the expense of another organism is a ___Parasite___.

2. Ticks and lice, since they live on the surface of a host act as ___Ectoparasites___.

3. Parasites that normally are free-living but can obtain nutrients from a host are ___Facultative___ parasites.

4. If a host harbors a parasite while it reproduces sexually it is a ___Definitive___ host.

5. Vectors which are required as part of the life cycle of the parasite are ___biological___ vectors.

6. Worms are ___Hermaphroditic___ if both male and female organs are found together.

7. Fungus-like protists include organisms called ___Slime___ molds.

8. Ciliates and amoebas are protists that are _animal-like_.

9. The red tide is caused by plant-like protists called _____.

10. Protozoans that move by pseudopodia are grouped in the _Sarcodinas_.

11. The presence of a complex-like cycle including merozoites and trophozoites would be found with the _____.

12. A loosely organized mass of thread-like structures that forms the thallus of most fungi is called the _____.

13. Cross walls found in the hyphae of most fungi are called _____.

14. Fungi with single septal pores often have organelles which act in protection called _____ _____.

15. Fungi that act as decomposers of organic matter are commonly called _____ fungi.

16. When fungi cause diseases in humans, the diseases are known as _____.

17. The ability of a fungus to exhibit different structural forms when it changes habitats is called _Dimorphisms_.

18. The bread molds such as *Rhizopus* are characterized by forming _____.

19. The sac fungi are so called because of the presence of sac-like _____.

20. The majority of the mushrooms and toadstools are found with the club fungi or _____.

21. The "imperfect" fungi are so-called because they lack a _____ _____.

22. Parasitic worms that look much like a measuring tape are called _____.

23. The head end of a tapeworm that is equipped with hooks and suckers is the _____.

24. Tapeworm larva that form into sac-like, bladder worms in human tissues are called _____.

25. Flukes can give rise to aquatic free-swimming larva called _____.

26. Body units of tapeworms that contain reproductive organs are called _____.

27. Free-swimming larval forms of flukes that penetrate snails or other mollusks are called _____.

28. Worms that enter the human body as immature larvae usually by means of arthropod vectors are _____.

29. Fungi are studied in the specialty called _____.

30. In the reproductive cycle of ascomycetes, the cytoplasmic fusion of the male and female strains is called _____.

Match the following terms with the description listed below.

Key Choices:

a.	Arachnids	e.	Flukes
b.	Insects	f.	Filarial worms
c.	Crustacea	g.	Helminths
d.	Tapeworms	h.	Nematodes

_____ 1. Arthropods with eight legs; examples: ticks and mites.

_____ 2. Some of these parasites can form hydatid cysts.

_____ 3. The general term for all the parasitic worms.

_____ 4. Often involves snails or other mollusks in their life cycle.

_____ 5. Includes the lice, fleas, and mosquitoes.

Match the following terms with the description listed below.

Key Choices:

a.	Water molds	e.	Imperfect fungi
b.	Bread molds	f.	Plant-like protists
c.	Sac fungi	g.	Animal-like protists
d.	Club fungi	h.	Fungal-like protists

_____ 6. Includes the diatoms and dinoflagellates.

_____ 7. Form pseudoplasmodia which are slightly motile cells and they are also saprophytes.

_____ 8. Produce motile spores that are usually found in water and include the mildews.

_____ 9. The poisonous mushrooms and organisms that produce ergot.

_____ 10. Includes yeasts such as *Candida* and *Saccharomyces*.

Match the following terms with the description listed below.

Key Choices:

a. Ectoparasites   f. Obligate parasites
b. Mechanical vectors   g. Accidental parasites
c. Definitive hosts   h. Biological vectors
d. Intermediate hosts   i. Permanent parasites
e. Endoparasites

_____ 11. Protozoa and worms that live within the bodies of other organisms.

_____ 12. Flies and cockroaches that carry parasite eggs or bacteria.

_____ 13. Ticks that ordinarily attach to dogs or other animals but sometimes attach to humans.

_____ 14. Hosts that harbor parasites during a stage other than a sexual one.

_____ 15. Tapeworms that remain in the host for an indefinite period of time.

REVIEW QUESTIONS

True/False (Mark T for True, F for False)

_____ 1. Plasmogamy is a process that is associated with the sexual reproduction of certain types of fungi.

_____ 2. Fungi are generally classified on the basis of their type of asexual reproduction.

_____ 3. Parasites that can effectively kill their host in a short period of time are those that have maintained the longest relationships with their hosts.

_____ 4. All animal protists are parasitic.

_____ 5. Flatworms lack a coelom and have a simple digestive tract with a single opening.

_____ 6. Most roundworms that parasitize humans live most of their life cycle in the circulatory system.

Multiple Choice

_____ 7. Conidia are associated with:
a. Club fungi
b. Plant-like protists
c. Sporozoites
d. Ascomycetes

_____ 8. Nematodes are characterized by:
a. having a pseudocoelom and separate sexes.
b. possessing a scolex.
c. forming woronin bodies.
d. having life cycles involving snails.

_____ 9. Hydatid cysts:
a. are formed by the sac fungi.
b. contain tapeworm larvae.
c. can be called ectoparasites.
d. usually exhibit dimorphism.

_____ 10. Arachnids:
a. include the crabs and crayfish.
b. have six legs.
c. are not known to transmit disease.
d. are vectors of Rocky Mountain Spotted Fever and scrub typhus.

ANSWER KEY

Fill-in-the-Blank

1. Parasite 2. Ectoparasites 3. Facultative 4. Definitive
5. Biological 6. Hermaphroditic 7. Slime 8. Animal-like
9. Dinoflagellates 10. Sarcodinas 11. Sporozoites 12. Mycelium
13. Septa 14. Woronin body 15. Saprophytic 16. Mycoses
17. Dimorphism 18. Zygospores 19. Asci 20. Basidiomycetes
21. Sexual stage 22. Tapeworms 23. Scolex 24. Cysticerci
25. Cercaria 26. Proglottids 27. Miracidia 28. Microfilaria
29. Mycology 30. Plasmogamy

Matching

1.a 2.d 3.g 4.e 5.b

6.f 7.h 8.a 9.d 10.c

11.e 12.b 13.g 14.d 15.i

Review Answers

1. **True** Plasmogamy is the process whereby haploid gametes of fungi can unite and allow their cytoplasm to fuse, ultimately forming a dikaryotic or a diploid cell. (p. 301; Figure 12.7)

2. **False** Fungi are actually classified on the basis of their sexual spore formation. Asexual reproduction is a very common method of propagation. (p. 304; Table 12.2)

3. **False** An effective, efficient parasite ensures continued survival by not killing its host. A parasite that has only a brief experience with a host will often kill it quickly because it lacks an appropriate tolerance. (pp. 294-295)

4. **False** Most animal protists are free-living and are often found in fresh or salt water. Only a few are actually parasites. (p. 298; Table 12.1)

5. **True** Flatworms are primitive worms usually no more than 1 mm thick with lengths of over 30 feet. Roundworms have a pseudocoelom and a complete digestive system. (p. 308)

6. **False** Most roundworm infestations in the United States affect the gastrointestinal tract. Therefore the majority of their life cycle is spent in the digestive tract. (p. 311)

7. **d** Conidia are asexual reproductive structures commonly seen with members of the sac fungi in the class Ascomycota. (p. 305)

8. **a** Tapeworms possess a scolex, flukes commonly have life cycles involving snails, and certain fungi have an organelle called a woronin body. (p. 308)

9. **b** Hydatid cysts are formed by certain types of small tapeworms of the genus *Echinococcus*. (p. 311)

10. **d** The arachnids include ticks, mites, spiders, and scorpions and have 8 legs and mouth parts for capturing and tearing apart prey. (p. 315; Table 12.4)

CHAPTER 13

STERILIZATION AND DISINFECTION

Soon after the discovery that microbes could cause disease, major efforts were made to control them. The application of heat as an effective control measure was certainly among the first significant methods to be developed. This was followed with the discovery by Semmelweiz and Lister that chemicals, specifically carbolic acid, could be used as effective control agents in terms of the transmission of these infectious agents. Eventually, a whole spectrum of chemicals, substances, and techniques was discovered and employed in the quest to control and/or eliminate microbes.

Although heat as a sterilizing agent still proves to be an effective method, other techniques such as membrane filtration, gamma radiation, and ultrasound are being relied upon as control methods for substances that cannot be sterilized by heat. New chemicals are being studied for their inhibitory properties as pathogens are becoming more and more resistant to the substances that initially proved to be so successful.

This chapter provides a description of the basic terms and the important principles of sterilization and disinfection followed by a survey of the most common types of chemical and physical control agents. Read the essay at the end of the chapter to appreciate the continued problems encountered by hospital personnel in terms of disease control and the need to search for new and even more successful control agents.

STUDY OUTLINE

I. **Definitions**
   A. Sterilization
   B. Disinfection
   C. Related terms

II. **Principles of Sterilization and Disinfection**
   A. Effects of heat
   B. Principles
   C. Application of principles

III. **Antimicrobial Chemical Agents**
   A. Potency of chemical agents
   B. Evaluation of effectiveness of chemical agents
   C. Disinfectant selection
   D. Mechanisms of action of chemical agents
   E. Specific antimicrobial chemical agents

IV. **Antimicrobial Physical Agents**
   A. Dry heat, moist heat, and pasteurization
   B. Refrigeration, freezing, drying, and freeze-drying
   C. Radiation
   D. Sonic and ultrasonic waves
   E. Filtration
   F. Osmotic pressure

REVIEW NOTES

A. How do sterilization and disinfection differ, and what terms are used to describe these processes?

Sterilization refers to the absolute killing or removal of all organisms in any material or on any object. There is no such thing as "almost" sterile. Disinfection refers to the reduction in numbers of pathogenic organisms on objects or in materials so that the organisms no longer pose a disease threat. Generally, disinfectants can not be used on the human body. Antisepsis refers to those agents that can be safely used on the body surfaces. See Table 13.1 for a list of all those terms commonly applied to sterilization and disinfection.

B. What important principles apply to the processes of sterilization and disinfection?

Microorganisms follow the same laws regarding death rates as those declining in numbers from natural causes. Therefore a definite proportion of organisms will die in a given time interval and will follow a logarithmic death rate. The smaller the number of organisms that are present, the less time will be needed to achieve sterility.

Based on these principles, thermal death points and thermal

death times have been devised for virtually all types of chemicals and equipment. It should be remembered that microbes will differ greatly in their susceptibility to antimicrobial agents, especially if they produce endospores.

TABLE 13.1 Terms related to sterilization and disinfection

Term	Definition
Sterilization	Killing or removing all microorganisms in a material or on an object.
Disinfection	Reducing the number of pathogenic microorganisms to the point where they pose no danger of disease.
Antiseptic	A chemical agent that can be safely used externally to destroy microorganisms or to inhibit their growth.
Disinfectant	A chemical agent used on inanimate objects to destroy microorganisms. Most disinfectants do not kill spores.
Sanitizer	A chemical agent typically used on food-handling equipment and eating utensils to reduce bacterial numbers so as to meet public health standards. Sanitization may simply refer to thorough washing with only soap or detergent.
Bacteriostatic agent	An agent that inhibits the growth of bacteria.
Germicide	An agent capable of killing microbes rapidly; some such agents are germicidal for certain microorganisms but only inhibitors for others.
Bactericide	An agent that kills bacteria. Most such agents do not kill spores.
Viricide	An agent that inactivates viruses.
Fungicide	An agent that kills fungi.
Sporocide	An agent that kills bacterial endospores or fungal spores.

C. What factors affect the potency of antimicrobial chemical agents?

Various factors affect the activity of antimicrobial chemical agents. These include the time used for exposure to the agent, the temperature of the agent, the pH, and the concentration of the agent. Potency increases with the length of time the organisms are exposed to the agent; to an increase in temperature; to an acidic or alkaline pH; and, to an extent, to an increase in the concentration of the agent. It is always extremely important to read the label to know exactly how to prepare the agent for the greatest effectiveness.

D. How is the effectiveness of an antimicrobial chemical agent assessed?

119

Effective evaluation of antimicrobial chemical agents is extremely difficult. No completely satisfactory method is currently available. What works in laboratory conditions may not work in clinical conditions. The most common evaluation technique is the phenol coefficient which can only be used for those disinfectants similar to phenol. The test involves the determination of ratios between phenol and the test disinfectant. Other tests include the filter paper method of evaluating chemical agents (see Figure 13.1) using small filter paper disks soaked with different chemical agents, and the use-dilution test where bacteria are added to tubes containing different dilutions of the agent.

E.   By what mechanisms do antimicrobial chemical agents act?

Antimicrobial chemical agents kill microbes by participating in one or more chemical reactions that damage the proteins, cell membranes, or other cell components.

Reactions that affect cellular proteins include hydrolysis, oxidation, and attachment of atoms or chemical groups to protein molecules. The reaction causes the protein to be denatured and nonfunctional. Agents include hydrogen peroxide, boric acid, halogens, heavy metals, and formaldehyde.

Reactions that affect cell membranes are surfactants that reduce the surface tension and dissolve lipids and protein denaturing agents such as those mentioned above. Surfactant agents include soaps and detergents, alcohols, phenols, and quaternary ammonium compounds.

Reactions of other chemical agents damage nucleic acids and energy-capturing systems. Damage to nucleic acids is an important means of inactivating viruses. Examples include alkylating agents such as ethylene oxide and nitrous acid, dyes, detergents, and alcohols.

F.   What are the properties of commonly used antimicrobial chemical agents?

The properties of antimicrobial chemical agents are summarized in **Table 13.2** which is illustrated on the next page.

G.   How are dry heat, moist heat, and pasteurization used to control microorganisms?

Table 13.4 provides an excellent summary of the properties of these antimicrobial agents. Use this table as a study guide when reviewing these mechanisms.

Heat destroys microorganisms by denaturing proteins, melting lipids, and when open flame is used, by incineration. Dry heat can be used to sterilize metal objects, oils, powders, and glassware but no liquid-containing material. A temperature of 160°C is used for 2 hours or longer.

Moist heat, especially when placed under pressure as with an autoclave, is the most effective method for sterilization. A temperature of 121°C is used for 15 minutes or longer. Heat-sensitive materials cannot be autoclaved. Quality control is essential when using autoclaves. Devices include heat-sensitive autoclave tape, wax pellets, and vials and strips containing endospores. The latter is the only device that can detect sterility.

Pasteurization does not achieve sterility. It does kill important pathogens such as those that cause tuberculosis, salmonellosis, brucellosis, Q-fever, and listeriosis.

TABLE 13.2 Properties of antimicrobial chemical agents

Agent	Actions	Uses
Soaps and detergents	Lower surface tension, make microbes accessible to other agents	Hand washing, laundering, sanitizing kitchen and dairy equipment.
Surfactants	Dissolve lipids, disrupt cell membranes, denature proteins, and inactivate enzymes in high concentrations; act as wetting agents in low concentrations	Cationic detergents are used to sanitize utensils; anionic detergents to launder clothes and clean household objects; quaternary ammonium compounds are sometimes used as antiseptics on skin.
Acids	Lower pH and denature proteins	Food preservation.
Alkalies	Raise pH and denature proteins	Found in soaps.
Heavy metals	Denature proteins	Silver nitrate is used to prevent gonococcal infections, mercury compounds to disinfect skin and inanimate objects, copper to inhibit algal growth, and selenium to inhibit fungal growth.
Halogens	Oxidize cell components in absence of organic matter	Chlorine is used to kill pathogens in water and to disinfect utensils; iodine compounds are used as skin antiseptics.
Alcohols	Denature proteins when mixed with water	Isopropyl alcohol is used to disinfect skin; ethylene glycol and propylene glycol can be used in aerosols.
Phenols	Disrupt cell membranes, denature proteins, and inactivate enzymes; not impaired by organic matter	Phenol is used to disinfect surfaces and destroy discarded cultures; amylphenol destroys vegetative organisms and inactivates viruses on skin and inanimate objects; chlorhexidine gluconate is especially effective as a surgical scrub.
Oxidizing agents	Disrupt disulfide bonds	Hydrogen peroxide is used to clean puncture wounds, potassium permanganate to disinfect instruments.
Alkylating agents	Disrupt structure of proteins and nucleic acids	Formaldehyde is used to inactivate viruses without destroying antigenic properties, glutaraldehyde to sterilize equipment, beta-propiolactone to destroy hepatitis viruses, ethylene oxide to sterilize inanimate objects that would be harmed by high temperatures.
Dyes	May interfere with replication or block cell wall synthesis	Acridine is used to clean wounds, crystal violet to treat some protozoan and fungal infections.

H. How are refrigeration, freezing, drying, and freeze-drying used to control and to preserve microorganisms?

Table 13.4 provides an excellent summary of the properties of these antimicrobial agents. All of these mechanisms can be used to retard the growth of microorganisms. Lyophilization, which is drying in the frozen state, can be used for long-term preservation of live cultures of microorganisms.

I. How are radiation, sonic and ultrasonic waves, filtration, and osmotic pressure used to control microorganisms?

Table 13.4 provides an excellent summary of the properties of these antimicrobial agents.
Radiation affects the DNA of microbes by damaging the nucleotides. Ultraviolet light, ionizing radiation, microwave radiation, and strong visible light can be used to control microorganisms and preserve foods.
Sonic and ultrasonic waves can kill microorganisms but they are used mostly for disruption of cells called sonication.
Filtration as a physical separation mechanism can be used to sterilize heat-sensitive substances, separate viruses, and collect microorganisms from air and water samples. Filters can be made of glass, asbestos, diatomaceous earth, and porcelain. The most common type used today is the membrane filter which is made of nitrocellulose.
Osmotic pressure can be created using high concentrations of sugar or salt that will plasmolyze the cells. It can be used to prevent the growth of microorganisms in highly sweetened or salted foods.

LEARNING ACTIVITIES

Complete each of the following items by supplying the appropriate word or phrase.

1. The killing or removal of all microorganisms in a material or on an object is called _Sterilization_ .

2. A chemical agent used to destroy most microbes on inanimate objects is a _Desinfectant_ .

3. An agent that can kill viruses would be called a _____ .

4. The time required to kill all bacteria in a particular culture at a specified temperature is the _thermal_ _death time_ .

5. The length of time needed to kill 90% of the organisms in a given population at a specified temperature is the _D value_ .

6. Disinfectants that are compared to phenol under the same conditions are employing a test called the _____ _____ _____.

7. When paper disks are soaked in selected disinfectants and placed on a plate seeded with a test organism, the test is called the _____ _____ _____.

8. Disinfectants that reduce the surface tension to break up grease particles are called _____.

9. Solutions that assist in the penetration of the disinfectant are called _____ _____.

10. Cationic agents that can be neutralized by soaps are _____ _____ compounds.

11. Pennies, nickels, and old silver dimes can act as inhibitory agents because they release tiny quantities of _Heavy_ _metals_.

12. Iodine combined with organic molecules used in skin preparation agents especially in surgery are _Iodophores_.

13. Chemicals dissolved in alcohols are called _Tinctures._.

14. Skin disinfectants that were found to be absorbed through the skin resulting in serious injury and death to dozens of babies in nurseries contained the chemical _____.

15. Hydrogen peroxide acts as an _oxidizing_ agent to disrupt disulfide bonds.

16. A gas that is commonly used to sterilize rubber and plastic tubing is _____ _____.

17. A device that sterilizes objects by moist steam under pressure is the _Autoclave_.

18. The temperature required for sterilization in a typical autoclave is _160_ °C.

19. A quality control device that can determine if an autoclave has actually sterilized a load of material contains _____.

20. A technique used to destroy pathogens in food products such as milk and wine is _Pasteurization_.

21. The process used in freeze-drying bacterial cultures and in making instant coffee is known as _____.

22. A unit of radiation energy absorbed per gram of tissue is a
_____ unit.

23. Disruption of cells by sound waves is called
_____ .

24. A pasteurization technique called _____ _____
_____ raises the temperature of liquids from
74°C to 148°C and back in less than five seconds.

25. A special type of filter called a _____
_____ _____ is used in surgical rooms and burn
units that require complete microbial control of the air.

Match the following terms with the description listed below.

   Key Choices:

   a. Pasteurization   f. Freezing
   b. Dry heat   g. Ultrasonic waves
   c. Moist, pressurized steam h. Freeze-drying
   d. Radiation   i. Osmotic pressure
   e. Filtration   j. Ethylene oxide

_____ 1. Used in sterilizing most laboratory culture media and
         hospital equipment.

_____ 2. Used to sterilize respiratory equipment but requires
         prolonged aeration.

_____ 3. Denatures proteins and damages nucleic acids by forming
         thymine dimers or breaking bonds.

_____ 4. Used to sterilize heat-sensitive solutions such as
         pharmaceuticals, vaccines, and water.

_____ 5. Causes plasmolysis of microbes.

_____ 6. Causes cavitation.

Match the following terms with the description listed below.

   Key Choices:

   a. Surfactants   f. Alkylating agents
   b. Halogens   g. Soaps and detergents
   c. Alcohols   h. Alkalies
   d. Phenols   i. Dyes
   e. Oxidizing agents

_____ 7. A substance that disrupts disulfide bonds.

124

_____ 8.   A substance found in soaps that raises the pH.

_____ 9.   A substance that denatures proteins and dissolves lipids especially when mixed with water.

_____ 10.   Disrupts cell membranes, denatures proteins, can be mixed with soap, and is the most common laboratory disinfectant.

_____ 11.   Includes formaldehyde and glutaraldehyde.

REVIEW QUESTIONS

True/False (Mark T for True, F for False)

_____ 1.   Chemical disinfectants are usually adversely affected by organic material.

_____ 2.   If a label on a disinfectant indicates it is bactericidal, it means it can kill any microbe and should be an effective sterilizing agent.

_____ 3.   If 20% of the organisms in an antimicrobial agent die in the first minute, than the rest should die within the next four minutes.

_____ 4.   It is virtually impossible to find a perfect disinfectant.

_____ 5.   Effective hand washing with bar soap can eventually sterilize the hands since most soaps kill bacteria on contact.

_____ 6.   Pasteurization does not achieve sterility but does kill pathogens present in the product.

Multiple Choice

_____ 7.   Select the most incorrect statement relative to sterilization principles.
a. A definite proportion of organisms will die in a given time interval.
b. Microorganisms tend to be similar in their susceptibility to antimicrobial agents.
c. The smaller the number of organisms present the shorter the time needed to achieve sterility.
d. If an agent cannot kill endospores it cannot be a sterilizing agent.
e. All of the above are correct.

_____ 8.   Select the most appropriate setting for sterilization.
           a.   Dry heat oven at 130°C for 1 hour.
           b.   Moist steam at 10 pounds of pressure for 15 minutes.
           c.   Moist steam at 121°C for 20 minutes.
           d.   Ethylene oxide at 20°C for 25 minutes.
           e.   Freezing at -10°C for 2 days.

_____ 9.   Microwave ovens:
           a.   cook by means of UV radiation.
           b.   cannot be used for sterilization because of uneven
                distribution of high-frequency vibrations.
           c.   can easily kill most bacteria and parasitic cysts.
           d.   actually cook by pasteurization.

_____ 10. The phenol coefficient test:
           a.   can be applied to all disinfectants.
           b.   uses *Escherichia coli* and *Bacillus subtilis* as
                indicator organisms.
           c.   compares the effectiveness of diluted phenol samples
                to dilutions of test samples.
           d.   is no longer used for testing purposes.

ANSWER KEY

Fill-in-the-Blank

1. Sterilization   2. Disinfectant   3. Viricide
4. Thermal death time   5. D value   6. Phenol coefficient test
7. Filter paper method   8. Surfactants   9. Wetting agents
10. Quaternary ammonium   11. Heavy metals   12. Iodophors
13. Tinctures   14. Hexachlorophene   15. Oxidizing
16. Ethylene oxide   17. Autoclave   18. 121°C   19. Endospores
20. Pasteurization   21. Lyophilization   22. Rad   23. Sonication
24. Ultra high temperature   25. High-efficiency particulate air

Matching

1.c 2.j 3.d 4.e 5.i 6.g

7.e 8.h 9.c 10.d 11.f

Review Answers

1.   **True** Most chemical disinfectants are unable to penetrate oily
or organic material. The addition of alcohol which can dissolve
oils often facilitates its action. (p. 325)

2.   **False** A bactericidal agent is one that kills bacteria but
often cannot kill bacterial endospores or viruses.
(p. 324, Table 13.1)

3.   **False**   The death rate of organisms follows a logarithmic scale, therefore, if 20% die in the first minute, 20% of the remaining organisms die in the next minute and so forth. At this rate, it may require 10 to 15 minutes to achieve sterility. (pp. 324-325)

4.   **True**   An ideal disinfectant should be fast acting, effective, penetrating, easily prepared, inexpensive, and safe for human use. No disinfectant currently available can accomplish all of these. (p. 326)

5.   **False**   Most hand soaps do not kill organisms but only act to remove surface bacteria, oily substances, and dirt. Because of the pores and crevices in the skin, it is impossible to sterilize the skin. (p. 328)

6.   **True**   Pasteurization is effective in killing organisms such as *Salmonella*, *Mycobacteria*, *Listeria*, Q-fever rickettsia, and *Brucella*. (p. 337)

7.   **b**   Microbes vary considerably in their susceptibility to an antimicrobial agent.   As a result it is necessary to perform susceptibility tests on these organisms to select the correct agent. (p. 325)

8.   **c**   The steam autoclave is normally operated at 121°C for 15-20 minutes depending on the size and type of material being sterilized. (p. 334)

9.   **b**   Microwave ovens produce high-frequency vibrations to cook foods by vibrating water molecules within the food. This in turn creates heat which is often unevenly distributed. (p. 340)

10. **c**   The phenol coefficient test cannot be applied to all disinfectants because of variations in the mechanism of action. The test uses *Staphylococcus aureus* and *Salmonella typhimurium*. *Bacillus subtilis* would never work because it produces endospores. The test is still being used today but on a limited scale. (p. 326)

CHAPTER 14

ANTIMICROBIAL THERAPY

One of the greatest achievements of this century was the discovery of antibiotics. Prior to that time as noted by the medical writer, Lewis Thomas, there was little chance of treating any of the serious microbial diseases.

The first major attempt to find specific chemical substances to treat infectious disease was made by Paul Ehrlich; however, with the discovery of penicillin by Fleming, the era of microbial control was at hand. New drugs were discovered and new chemicals synthesized. However, with every new conquest there is always some danger. The emergence of antibiotic-resistant strains of bacteria has posed major problems in the treatment and control of many disease agents. Today, we are barely a step ahead. Hopefully, with new discoveries and a better understanding of the genetics involved with microbial resistance, the future will again look promising.

The first part of this chapter presents the exciting development of antimicrobial therapy. This is followed with an interesting discussion of the general properties of these agents and how microbes have developed resistance to them. Determination of microbial sensitivities to these agents is also presented.

In the second part, a complete description of the most important antimicrobial agents including those that affect the pathogenic fungi, viruses, protozoans, and helminths is provided. The chapter ends with a discussion of some of the problems that are presented to hospitals in their control of the newly emerging resistant strains of bacteria.

STUDY OUTLINE

I. **Antimicrobial Chemotherapy**
    A. General features
    B. Terms
II. **History of Chemotherapy**
    A. Early history
    B. Ehrlich's contributions
    C. Development of sulfa drugs
    D. Fleming's discoveries
III. **General Properties of Antimicrobial Agents**
    A. Selective toxicity
    B. Spectrum of activity
    C. Modes of action
    D. Kinds of side effects
    E. Resistance of microorganisms
IV. **Determination of Microbial Sensitivities to Antimicrobial Agents**
    A. Disk diffusion method
    B. Dilution methods
    C. Serum killing power
    D. Automated methods
V. **Attributes of an Ideal Antimicrobial Agent**
VI. **Antibacterial Agents**
    A. Sources of antibiotics
    B. Inhibitors of cell wall synthesis
    C. Disrupters of cell membranes
    D. Inhibitors of protein synthesis
    E. Inhibitors of nucleic acid synthesis
    F. Other antibacterial agents
VII. **Other Antimicrobial Agents**
    A. Antifungal agents
    B. Antiviral agents
    C. Antiprotozoan agents
    D. Antihelminthic agents
    E. Special problems with resistant hospital infections

REVIEW NOTES

A. What terms are used to discuss chemotherapy and antibiotics, and what do they mean?

Chemotherapy refers to the use of any chemical agent in the practice of medicine. Chemotherapeutic agents are any chemicals used in medical practice while antimicrobial agents are chemicals used to treat diseases caused by microbes. A chemosynthetic agent is one that is made in the laboratory while an antibiotic agent is one that is produced by other microorganisms. Many of the drugs available today are semisynthetic in that they are partly made by microorganisms and partly by laboratory synthesis.

B.   How have chemotherapeutic agents been developed?

Chemotherapeutic agents were used for thousands of years and most were extracts from plants and herbs. The first systematic attempt to find chemotherapeutic agents was begun by Ehrlich. His work helped in the discovery of agents such as sulfa drugs. Fleming was the first to discover the antibiotics with his work on penicillin.

C.   How do the terms selective toxicity, spectrum of activity, and modes of action apply to antimicrobial agents?

Antimicrobial agents share certain common properties. These include selective toxicity, spectrum of activity, and mode of action. Selective toxicity refers to the property of antimicrobial agents that allows them to exert greater toxic effects on microbes than on the host. Some drugs such as penicillin have a very wide range between the level sufficient to affect the pathogen and the level of toxicity to the host. Others such as those that contain arsenic must be monitored very carefully because of their narrow range.

Spectrum of activity refers to the variety of microbes sensitive to the agent. Agents such as ampicillin and gentamicin that affect a large number of organisms are called broad spectrum. Agents such as erythromycin and penicillin that affect only a few are called narrow spectrum. See Table 14.1 for a list of drugs based on their spectrum of activity.

Mode of action refers to the mechanism by which the antibiotic exerts its effect. Some affect only the cell wall, some the cell membrane and some the cellular machinery or its genetic mechanisms. See Figure 14.2 on the next page which illustrates the major modes of action of selected antimicrobial agents.

D.   What kinds of side effects are associated with antimicrobial agents?

The side effects caused by antimicrobial agents include excessive toxicity, allergy, and disruption of normal flora bacteria. Many agents are not licensed for use because of their high toxicity to the patient even though they are very effective antimicrobial agents. Some drugs cause allergic reactions which represent a reaction to the agent as a foreign substance. Many drugs attack not only the pathogen but the normal flora that in many cases help prevent secondary infections. Elimination of these bacteria can cause superinfections with new pathogens.

E.   What is resistance to antibiotics, and how do microorganisms acquire it?

Resistance means that a microorganism formerly susceptible to an antibiotic is no longer affected by it. Nongenetic resistance occurs when microbes hide from the effects of the drug or undergo a temporary change that allows them to resist it.

Genetic resistance occurs when microbes are able to acquire new genetic units such as plasmids which carry resistant genes or when the microbes can alter their genes to avoid damage by the antibiotic. Chromosomal resistance is due to a mutation in the microbial DNA while extrachromosomal resistance is due to the acquisition of plasmids. Five mechanisms of resistance have been identified and include alteration of receptors, alteration of cell membrane permeability, development of new enzymes, alteration of existing enzymes, and alteration of metabolic pathways.

Drug resistance can be minimized by continuing the treatment until all sensitive pathogens are eliminated, combining the effectiveness of two antibiotics that do not have cross-resistance and by using antibiotics only when absolutely necessary.

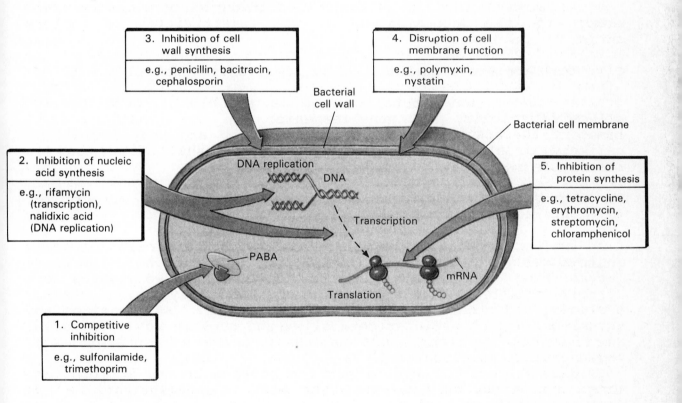

Figure 14.2: Modes of action of antibiotics.

131

F. **How are sensitivities of microbes to chemotherapeutic agents determined?**

Microbes vary greatly in their sensitivities to different chemotherapeutic agents, and their susceptibilities can change over time. By exposing the microbes to various agents in the laboratory a complete pattern of sensitivity or resistance can be determined. Several methods are available and include disk diffusion (Kirby-Bauer) method, dilution method, and the automated method.

The disk-diffusion method involves placing antibiotic-soaked paper disks on plates seeded with the test organism. Following incubation, zone diameters indicating the degree of sensitivity can be measured.

The dilution method involves the addition of an organism into tubes or wells containing nutrient media and varying concentrations of the antibiotic. A minimum inhibitory concentration of the agent can be determined and represents the lowest concentration in which no growth of the organism is observed.

Automated methods allow rapid identification of microorganisms along with the determination of its sensitivities to various antimicrobial agents.

G. **What are the attributes of an ideal antimicrobial agent?**

An ideal antimicrobial agent is soluble in body fluids, selectively toxic, and nonallergenic; can be maintained at a constant therapeutic concentration in blood and body fluids; is unlikely to elicit resistance; has a long shelf life; and is reasonable in cost.

H. **What are the properties, uses, and side effects of antibacterial agents?**

Antibacterial agents such as penicillin and cephalosporins inhibit cell wall synthesis. See Figure 14.10 and note the beta-lactam ring found with these antibiotics. Polymyxins and tyrocidins affect the cell membrane. Aminoglycosides, tetracyclines, and others affect protein synthesis. Sulfonamides and quinolones affect cellular metabolism and nucleic acid synthesis. See Figure 14.13 which illustrates the structures and properties of these antibacterial agents.

Note Table 14.2 which lists the sources of the antibiotics. It's interesting to note that the most common antibiotics are derived from various species of fungi while the greatest variety of antibiotics are derived from species of _Streptomyces_.

I. **What are the properties, uses and side effects of antifungal agents, antiviral agents, antiprotozoan agents, and antihelminthic agents.**

Antifungal agents such as the imidazoles affect the cell wall of fungal cells. Amphotericin B and nystatin cause cell membrane

132

permeability. Griseofulvin and flucytosine impair synthesis of nucleic acids and affects cellular proteins. Figure 14.15 provides a summary of the structures and properties of antifungal agents.

Antiviral agents have been difficult to develop because they must damage viruses inside the cells without affecting the host cell. Antibiotics do not affect viruses. Most antiviral agents are analogs of purines or pyrimidines. These include idoxuridine, ribavirin, and acyclovir. Amantadine seems to prevent influenza A from penetrating cells. Interferon which is produced and released by infected cells to stimulate other cells to produce antiviral proteins has been of limited success in treating viral infections. See Figure 14.15 which summarizes the structures and properties of various antiviral drugs.

Antiprotozoan agents such as chloroquine and primaquine interfere with protein synthesis while others such as pyrimethamine affect folic acid synthesis. The mechanism of some agents is not well understood. See Figure 14.15 for a summary of their structures and properties.

Antihelminthic agents such as niclosamide and mebendazole affect carbohydrate metabolism of the worms while others such as piperazine act as powerful neurotoxins. See Figure 14.15 for a summary of their properties and structures.

J.   How do resistant hospital infections arise, and how can they be treated and prevented?

Soon after the advent of antibacterial agents, antibiotic-resistant organisms began to appear. Many infections caused by these resistant strains were due to the intensive or inappropriate use of the antibiotics. Since the hospital environment is conducive to the development of these strains and because the hospitalized patients often have a lowered resistance, these infections are more common and are usually severe. Treatment and prevention of these infections is extremely difficult. It is best to use antibiotics only when needed, to apply them aggressively, and to apply them appropriately.

LEARNING ACTIVITIES

Complete each of the following items by supplying the appropriate word or phrase.

1.   A chemical substance produced by one microbe that is inhibitory or lethal to another is an _____.

2.   Chemical agents produced in the laboratory that are used in therapy are _____ agents.

3.   Antimetabolites that imitate the normal molecule required for their metabolism are probably affecting it by _____ _____.

4. The use of chemicals to kill pathogenic organisms is known as
    _____.

5. Alexander Fleming is credited with discovering the first
    antibiotic known as _____.

6. Antimicrobial agents that can attack a wide variety of
    microbes can be said to be _____ _____.

7. Sulfa drugs affect microbes by acting as _____.

8. Penicillin contains a chemical structure called the _____
    _____ _____which acts as the active portion of the
    molecule.

9. Some antimicrobial agents that allow invasion by replacement
    flora tend to cause _____.

10. Extrachromosomal pieces of DNA that can carry resistance genes
    and can be transmitted from one cell to another are called ___
    _____.

11. A method used to determine antibiotic sensitivity which
    involves filter paper disks soaked in the antibiotic is the
    _____ _____.

12. Measurable clear areas around disks used in the determination
    of sensitivities to antibiotics in the Kirby-Bauer test are
    called _____ ___ _____.

13. If two separate antibiotics cannot inhibit an organism but
    together they can exert an effect, the phenomenon is known as
    _____.

14. The greatest variety of antibiotics are produced by members
    of the genus _____.

15. Cephalosporins are derived from species of _____.

16. Purine and pyrimidine analogs appear to be effective in
    controlling infections caused by _____.

17. A substance acquired from the bark of the chinchona tree that
    was used to treat malaria is _____.

18. Black hairy tongue sometimes occurs as a result of the
    antiprotozoan drug _____.

19. The maximum tolerable dose of an antibiotic per kilogram body
    weight devided by the minimum dose per kilogram body weight
    that will cure the disease is called its _____
    _____.

Match the following terms with the description listed below. Terms may be used more than once.

Key Choices:

a. Penicillins   f. Macrolides
b. Cephalosporins   g. Sulfonimides
c. Polymyxins   h. Isoniazid
d. Aminoglycosides   i. Rifampin
e. Tetracyclines

_____ 1.   Inhibits cell wall synthesis.

_____ 2.   Affects primarily the cell membrane.

_____ 3.   Acts as antimetabolites to bacterial cells.

_____ 4.   May cause color stains on newly forming teeth.

_____ 5.   Obtained from species of *Streptomyces* and are often used in conjunction with other drugs to increase their effectiveness.

_____ 6.   May cause "red man" syndrome.

_____ 7.   Major example is erythromycin which is the drug of choice for Legionnaires' disease.

Match the following terms with the description listed below. Terms may be used more than once.

Key Choices:

a. Amphotericin B   g. Interferon
b. Imidazole   h. Chloroquine
c. Nystatin   i. Metronidazole
d. Griseofulvin   j. Niclosamide
e. Amantadine   k. Mebendazole
f. Purine & pyrimidine analogs l. AZT

_____ 8.   Used primarily in the control of tapeworm infestations.

_____ 9.   May cause "black hairy tongue" when used to treat trichomonas infections.

_____ 10. A substance which induces the formation of antiviral proteins in uninfected cells.

_____ 11. Blocks the uptake of glucose by roundworms and pinworms.

_____ 12. Useful in the prevention of influenza A virus infections.

_____ 13. Primary drug of choice against most parasites of malaria.

_____ 14. The drug currently used to treat the viruses of AIDS.

_____ 15. Drug given IV to control serious systemic fungal infections.

_____ 16. Antifungal drug derived from <u>Penicillium</u> used to control infections of skin, hair, and nails.

REVIEW QUESTIONS

True/False (Mark T for True, F for False)

_____ 1. Antibiotics such as penicillin that can inhibit the formation of the cell wall have a bacteriocidal effect.

_____ 2. The use of antibiotics in animal food supplements has greatly reduced the number of infections transmitted in poultry, especially chickens and turkeys.

_____ 3. A major problem with overuse of clindamycin is that superinfections of *Clostridium difficile* could occur.

_____ 4. Certain cold viruses seem to be controlled best by antibiotics such as the aminoglycosides that affect their protein synthetic processes.

_____ 5. Once a patient feels better following a bacterial infection, antibiotic therapy should be halted to reduce the problems of overuse of antibiotics.

_____ 6. Tetracyclines could actually stain teeth if given prior to the eruption of the teeth.

Multiple Choice

_____ 7. Select an attribute of an antibiotic that would not be beneficial.
   a. Toxicity not easily altered
   b. Insolubility in body fluids
   c. Nonallergenic
   d. Maintenance of constant, therapeutic concentration in blood and tissues
   e. All of the above are beneficial

_____ 8. The minimal inhibitory concentration test:
   a. uses disks soaked in antibiotics.
   b. allows measurement of zones of inhibition.
   c. can help determine the appropriate dilution of an antibiotic for use in therapy.
   d. involves the use of the patient's serum.

_____ 9.   In respect to antibiotic therapy:
           a.   antibiotics should always be used to treat viral
                infections.
           b.   superinfections can be controlled by using large
                doses of tetracyclines.
           c.   because of the development of resistant strains some
                drugs have gone to second and third generations of
                derivatives.
           d.   drugs that affect protein synthesis will have the
                greatest bactericidal effect.

_____ 10. Antiprotozoan agents:
           a.   include metronidazole and chloroquine.
           b.   all are designed to affect the cell wall.
           c.   are laboratory-synthesized penicillin derivatives.
           d.   can also be used to treat viral agents.

_____ 11. Which of the following is not true in respect to
           antibiotic resistance in bacteria.
           a.   Chromosome resistance usually occurs against a single
                type of antibiotic.
           b.   Extrachromosomal resistance usually involves R
                factors.
           c.   Resistance can occur by altering the receptors to
                which antimicrobial agents bind.
           d.   Resistance can occur by altering a metabolic pathway
                which bypasses a reaction inhibited by an antimicro-
                bial agent.
           e.   All of above are true.

ANSWER KEY

Fill-in-the-Blank

1. Antibiotic 2. Synthetic 3. Molecular mimicry 4. Chemotherapy
5. Penicillin 6. Broad spectrum 7. Antimetabolites
8. Beta-lactam ring 9. Superinfections 10. R plasmids
11. Kirby-Bauer 12. Zones of inhibition 13. Synergism
14. Streptomyces 15. Fungi 16. Viruses 17. Quinine
18. Metronidazole 19. Chemotherapeutic index

Matching

1.a,b 2.c 3.i,g,h 4.e 5.d 6.i 7.f

8.j 9.i 10.g 11.k 12.e 13.h 14.l 15.c,a 16.d

Review Answers

1.   **True**   Antibiotics that inhibit the cell wall synthesis selectively damage bacterial cells. Since many bacteria have a very high internal osmotic pressure, the loss of a wall will cause the cells to burst when subjected to a low osmotic condition. (pp. 352-353)

2.   **False** Antibiotics used in animal feeds have greatly increased the resistant strains of many bacteria such as *Salmonella* and *Escherichia coli*. The low dosage levels in the products provide the opportunity for resistance to develop. (p. 358)

3.   **True** Clindamycin is effective against *Bacteroides* and other anaerobes but not *Clostridium difficile*. Toxins produced by this anaerobe can cause a severe and sometimes fatal colitis. (p. 366)

4.   **False** Since viruses reproduce only inside living cells, they do not respond to any antibiotics and only respond to just a few chemosynthetic agents. (p. 371)

5.   **False**   It is very important to finish all antibiotic medications to insure that all microbes are destroyed. Failure to do so may allow a few persisters to survive as well as a few resisters which could then establish a resistant population. (p. 357; Figure 14.6)

6.   **True** Tetracyclines, while effective against a wide variety of organisms, have a major side effect which is permanent staining of teeth.   The drug should be avoided by pregnant mothers and very young children. (pp. 364-365)

7.   **b**   Antibiotics should be soluble in body fluids such as blood and cerebrospinal fluid to insure proper distribution. (p. 361)

8.   **c**   Dilution methods provide a determination of the lowest concentration that will inhibit the growth of an organism. The disk diffusion method uses antibiotic-soaked disks which allows determination of sensitivities based on zone diameter measurements. (p. 360; Figure 14.7)

9.   **c**   Antibiotics are not effective against any viruses; tetracyclines often eliminate the normal flora and promote superinfections; and drugs affecting protein synthesis only exert a bacteriostatic effect. (pp. 352-354)

10. **a**   Most antiprotozoan agents affect the metabolism of the parasites. Protozoans also do not possess a cell wall that could be affected by drugs such as penicillin. Since the drugs affect metabolism, they would have no effect on viruses. (pp. 372-373)

11. **e**   All of the mechanisms do occur to provide microbes with resistance to antimicrobial agents. (pp. 356-357)

CHAPTER 15

HOST—MICROBE RELATIONSHIPS AND DISEASE PROCESSES

Microorganisms have established many important relationships with humans. As this chapter indicates several of these relationships such as mutualism and commensalism are extremely valuable and provide useful benefits to the host. In only a few instances have microbes found it necessary to turn on their host and cause disease. As these infectious agents developed their harmful, parasitic relationships most of them still found it to their benefit to keep the host alive and reasonably well, thus ensuring their own survival. In view of this, it's often useful to understand these relationships from the microbe's point of view as well as from the host's perspective. Once an organism establishes itself in a disease-causing relationship, the process can be monitored in terms of what the organism possesses or produces to cause disease, what signs or symptoms indicate infection, what types of infections they can cause, and what steps they normally follow in the disease process.

The chapter begins with a discussion of the basic types of host-microbe relationships. This is followed with a description of the types of diseases and a discussion of the disease process. The essay at the end of the chapter presents an interesting review of how we have waged war on infectious diseases. In some areas we have certainly won the battles, in others, the parasites still seem to have the upper hand.

STUDY OUTLINE

I.   **Host-Microbe Relationships**
    A.   Symbiosis
    B.   Contamination, infection, and disease
    C.   Pathogens, pathogenicity, and virulence
    D.   Normal flora
II. **Koch's Postulates**
    A.   The four postulates
    B.   Criteria for postulates
III. **Kinds of Disease**
    A.   Infectious versus noninfectious diseases
    B.   Classification of diseases
    C.   Communicable versus noncommunicable diseases
    D.   Exogenous versus endogenous diseases
IV. **The Disease Process**
    A.   How microbes cause disease
    B.   Signs, symptoms, and syndromes
    C.   Types of infectious diseases
    D.   Steps in the course of an infectious disease

REVIEW NOTES

A.   **What terms are used to define host-microbe relationships and what do they mean?**

Microorganisms exhibit a variety of complex relationships to other microbes and to the larger organisms that serve as hosts for them. A host is any organism such as a human or an animal that harbors another organism.

Symbiosis refers to any interactions between two organisms; mutualism is where both organisms benefit by the relationship; commensalism is where one organism benefits and the other is not affected; and parasitism is where one organism benefits and the other is harmed.

Various terms are used to illustrate the conditions in which microorganisms have increasingly significant effects on their hosts. Contamination refers to the mere presence of microbes, infection refers to the invasion of the host by the microbes (however, this invasion may go unnoticed, i.e., inapparent infection), and disease refers to the development of symptoms that indicate a disturbance in the health of the host.

A pathogen is a parasite that is capable of causing disease, pathogenicity is the capacity to produce disease, and virulence is the degree of intensity of a disease.

Normal flora represent organisms found normally on or in another organism. Resident flora are always present while transient flora are present under certain conditions. Opportunists are either resident or transient flora that can cause disease when the host has been compromised or weakened. See Table 15.1 which illustrates the normal flora found on or in the human body. Notice that *Staphylococcus* species are found in every body area. These organisms are very common endogenous infectious agents, and for good reason!

B.   How do Koch's postulates relate to infectious disease?

Koch's postulates are used to provide a link between an infectious organism and a disease. When all four of the postulates are met, an organism has been proven to be the causative agent of the disease. Note that the most important part of the postulates is the ability to pure-culture the organisms. Until that is done, there is no way to be absolutely sure that it is the causative agent. See Figure 15.3 for a demonstration of these postulates.

C.   What are the major differences between infectious and noninfectious disease, between communicable and noncommunicable infectious diseases, and between exogenous and endogenous diseases?

Infectious diseases are caused by infectious agents such as bacteria and viruses, while noninfectious diseases are caused by other factors such as genetic defects, nutritional deficiency defects, etc. Note that many infectious agents often interact with other factors to cause disease.

Communicable diseases such as diphtheria and whooping cough can be spread from one host to another, while noncommunicable infectious diseases such as tetanus cannot be spread from host to host and usually are acquired from the soil or water.

Exogenous diseases such as tuberculosis and plague are those caused by pathogens or other factors from outside the body, while endogenous diseases such as a urinary tract infection are caused by pathogens found on or in the body.

D.   How do microbes cause disease?

Microorganisms act in certain ways that allow them to cause disease. They can cause disease by adherence to a host, by colonization and/or invasion of the host tissues, and by invading the host cells.

Many organisms produce toxins which are synthesized inside the cell. Exotoxins are released from the pathogen, are generally very specific, and are highly toxic. Endotoxins are part of the outer portion of the cell wall of bacteria and are released only when the cell dies. See Table 15.2 for a list of the properties of toxins and Table 15.3 for examples of bacteria producing them. Note that many of the organisms that release exotoxins are Gram-positive whereas those that possess endotoxins are usually Gram-negative.

Bacteria can release other substances such as hemolysins which lyse red blood cells, leukocidins which kill white blood cells, hyaluronidases which allow them to spread throughout tissues, coagulases which form blood clots, and fibrinolysins which dissolve fibrin blood clots. It is important to remember that it is not the

141

number of toxins or virulence factors but the type of factors that determines the severity of the microbial attack.

Viruses damage cells and produce a cytopathic effect such as altering the DNA, forming inclusion bodies, or affecting the cell's protein synthesis machinery. The effect is cytocidal if the cell dies and noncytocidal if it does not.

Fungi can progressively digest cells and produce damaging toxins. Protozoa and helminths damage tissues by ingesting cells and tissue fluids, releasing toxic waste products, and by causing allergic reactions.

E. What are the meanings of terms used to describe diseases?

A sign is an observable effect of a disease and a symptom is an effect of a disease that is reported by the patient. A syndrome is a group of signs and symptoms that occur together. Since signs and symptoms are often similar with many diseases, this information is extremely valuable to a physician to determine the exact nature of the patient's illness. Tables 15.4 and 15.5 provide excellent summaries of the correlation of signs and symptoms with tissue damage and the terms used to describe infections.

F. What steps occur in the course of an infectious disease?

Most diseases caused by infectious agents have a fairly standard course or stages that can be monitored. These stages include incubation, prodromal, invasive, acme, decline, and convalescent. Even with treatment, the disease still passes through all of these stages.

The incubation period is the time between infection and the appearance of the signs and symptoms of the disease. See Figure 15.9 for the incubation periods of selected diseases. Note that some such as cholera and diphtheria are short whereas others such as rabies and hepatitis B are fairly long.

The prodromal phase is the period during which the pathogens begin to invade the tissues which is marked by early nonspecific symptoms such as a cough or sneeze.

The invasive phase is the period during which pathogens invade and significantly damage or affect tissues. This phase is marked by the most classic signs and symptoms of the disease.

The acme, or critical stage, is the period of most intense symptoms.

The decline phase is the period during which the host defenses overcome the pathogens, the signs and symptoms subside, and the body's immune defenses have surmounted the infectious agent.

The convalescent phase is the period when tissue damage is repaired and the patient begins to regain strength. Sometimes during this phase or shortly thereafter, latent effects of the disease called sequelae appear. These could include cranial nerve damage, heart valve damage, hearing loss, and others.

142

LEARNING ACTIVITIES

Complete each of the following items by supplying the appropriate word or phrase.

1. Two species living together and both benefit from the relationship is an example of _Mutualism_.

2. Any organism that harbors another organism is a _host_.

3. A symbiotic relationship where only one member benefits while the other is not harmed is _Commensalism_.

4. The multiplication within the body in an abnormal number or abnormal location by any parasitic organism is an _____.

5. Any parasite capable of causing disease in its host is a _Pathogen_.

6. The degree of intensity of a disease organism refers to its _Virulence_.

7. A pathogen's virulence can be decreased by _Attenuation_.

8. The organisms that normally reside on or in the host may be called _Normal_ _Flora_ or _Resident_.

9. Organisms present on or in the body for temporary periods of time are probably _Transient_.

10. Organism that cause disease by taking advantage of a reduction in the host's resistance are called _opportunist_.

11. Diseases caused by any factors other than infectious organisms are _non_ _infectious_.

12. Diseases due to errors in genetic information are called _inherited_ diseases.

13. Diseases actually caused by the medical treatment are _Iatrogenic_ diseases.

14. Diseases whose cause is unknown are _____.

15. Diseases caused by the ingestion of toxins accumulating in a product are _Intoxications_.

16. Diseases that are acquired from the environment and not spread from one host to another are _non communicable_ _infectious_ diseases.

17. Substances that allow bacteria to attach to host cells are called _____.

18. A soluble, poisonous substance secreted by bacteria into host tissues is an _____.

19. Toxins that act directly on tissues of the gut are _Exotoxins_____.

20. A toxin that is found mostly with Gram-negative organisms and is of a lipopolysaccharide construction is an _Endotoxins_.

21. Enzymes produced by certain bacteria that can kill white blood cells are _Leukocidins_____.

22. Hemolysins that completely lyse red blood cells are _Beta_ hemolysins.

23. An enzyme produced by certain bacteria that accelerates the coagulation of blood is a _coagulase_____.

24. Viruses that do not actually kill host cells may be called _____ viruses.

25. An _____ infection occurs when a virus invades a cell but is unable to make infectious progeny.

26. Species of *Giardia* possess a structure called _____ which allows them to attach to cells that line the gastrointestinal tract.

27. When a disease causes aftereffects, the condition is known as a _____.

28. A _____ disease develops slowly and persists for long periods of time.

29. A disease characterized by periods of inactivity is a _latent_____ disease.

30. An infection confined to a specific area but from which pathogens can spread is a _____ infection.

31. Pathogens that are present in the blood and are multiplying characterize a _Septicemia_____.

32. An infection that fails to produce symptoms can be an _subclinical_____ or _inapparent_____ infection.

33. The disease period during which nonspecific symptoms appear such as weakness and fever is the _prodromal_____ period.

144

34. The most severe signs and symptoms occur during the ___Invasive___ phase.

35. _____ are substances produced by bacteria that cause fever.

36. The recovery period of a disease is known as the ___Convalescence___ period.

37. When a disease reaches a period of most intense symptoms it is in the ___acme___ stage.

Match the following terms with the description listed below. Choices may be used more than once.

Key Choices:

a. Symbiosis   f. Competitive exclusion
b. Mutualism   g. Infestation
c. Commensalism   h. Infection
d. Contamination   i. Opportunists
e. Parasitism

___i___ 1. The hosts own bacteria that invade tissues following surgery.

_____ 2. Tapeworms that are living in the intestines of a cat.

_____ 3. The presence of staphylococcal bacteria on a soft drink can.

___?___ 4. Normal intestinal bacteria that prevent colonization by others.

___b___ 5. Intestinal bacteria that provide vitamins to the host and, in turn, obtain a living space.

Match the following terms with the description listed below. Choices may be used more than once.

Key Choices:

a. Communicable disease   f. Immunological disease
b. Exogenous disease   g. Degenerative disease
c. Endogenous disease   h. Noninfectious disease
d. Noncommunicable disease   i. Mental disease
e. Iatrogenic disease

_____ 6. Legionnaires' bacteria that originate from the soil and cause a respiratory infection.

_____ 7. A patient with bacterial endocarditis in which the disease progressively becomes worse over a period of time.

_____ 8. An unknown viral disease transmitted in uncooked pork that affects antibody producing B-cells.

_____ 9. A child with chicken pox.

_____ 10. A chronic disease such as syphilis that ultimately causes senility.

_____ 11. A malignancy in the bone marrow caused by excessive radiation exposure due to a poorly operating X-ray machine.

Match the following terms with the description listed below.

Key Choices:

a. Exotoxins   f. Beta hemolysins
b. Endotoxins   g. Hyaluronidases
c. Enterotoxins   h. Fibrinolysins
d. Toxoids   i. Coagulases
e. Alpha hemolysins

__a__ 12. A polypeptide poison produced by *Corynebacterium diphtheriae* that circulates in the blood and rapidly kills cells.

__e__ 13. An enzyme produced by certain streptococci that will partially lyse red blood cells.

__d__ 14. A type of vaccine used to prevent tetanus made by inactivating the poison it produces.

__h__ 15. A substance that dissolves blood clots.

__g__ 16. The spreading factor produced by certain bacteria.

__b__ 17. The cell wall of *Salmonella* bacteria.

Match the following terms with the description listed below.

Key Choices:

a. Acute disease   f. Secondary infection
b. Latent disease   g. Systemic infection
c. Chronic disease   h. Mixed infection
d. Viremia   i. Bacteremia
e. Superinfection

_____ 18. Presence of viruses in the blood.

_____ 19. The observation of rapidly developing symptoms.

_____ 20. An infection which follows an earlier one as a possible result of a weakened condition.

_____ 21. Destruction of the normal flora can often result in this condition.

_____ 22. When pathogens spread throughout the body especially through the lymphatics.

REVIEW QUESTIONS

True/False (Mark T for True, F for False)

_____ 1. The most successful parasites are those that successfully invade a host and quickly kill it.

_____ 2. Koch's postulates can be easily applied to all bacteria and even viruses since they can be cultivated artificially in the laboratory.

_____ 3. Some pathogens such as *Giardia* possess substances called adhesive disks that allow them to attach to specific tissue of the body.

_____ 4. Endotoxins, although poisonous, never cause the death of a host; however, exotoxins can.

_____ 5. Viruses that cause noncytopathic effects generally form inclusion bodies within cells before the cell is killed.

_____ 6. A latent disease may be characterized by periods of inactivity either before or after an attack.

_____ 7. Certain bacteria lyse blood cells because the cells contain magnesium which is essential for microbial replication.

Multiple Choice

_____ 8. An exotoxin can be characterized by all the following except:
a. found usually with Gram-positive and a few Gram-negative bacteria.
b. can easily be converted to a toxoid.
c. relatively unstable.
d. causes a rapid rise in fever.

147

_____ 9. If a hospitalized patient obtained an infection from a
nurse because of poor hand washing it could be:
a. an intoxication.
b. an idiopathic disease.
c. an endogenous disease.
d. an infestation.
e. none of the above are correct.

_____ 10. A substance released by a microbe that could cause it to
spread throughout tissues would be:
a. hyaluronidase   c. leukocidins
b. coagulase   d. hemolysins

_____ 11. All of the following are true except:
a. any parasite that can cause disease is a pathogen.
b. the degree of intensity of a disease is known as
virulence.
c. attenuation means that the disease-producing ability
of the organism has dramatically increased.
d. increasing animal passage can increase the virulence
of a pathogen.

ANSWER KEY

Fill-in-the-Blank

1. Mutualism 2. Host 3. Commensalism 4. Infection 5. Pathogen
6. Virulence 7. Attenuation 8. Normal flora, resident
9. Transients 10. Opportunists 11. Noninfectious diseases
12. Inherited 13. Iatrogenic 14. Idiopathic 15. Intoxications
16. Noncommunicable infectious disease 17. Adhesions 18. Exotoxin
19. Enterotoxins 20. Endotoxin 21. Leukocidins 22. Beta
23. Coagulase 24. Noncytocidal 25. Abortive 26. Adhesive disk
27. Sequelae 28. Chronic 29. Latent 30. Focal 31. Septicemia
32. Inapparent, subclinical 33. Prodromal 34. Invasive
35. Pyrogens 36. Convalescence 37. Acme

Matching

1.h,i 2.g,e 3.d 4.f 5.b,c

6.b,d 7.c,d,g 8.c,d,f 9.a,b 10.b,g,i 11.e,h

12.a 13.e 14.d 15.h 16.g 17.b

18.d 19.a 20.f 21.e 22.g

Review Answers

1. **False** A successful parasite is one that can ensure the
continued survival of its host without causing too much adverse
damage. (p. 386)

2. **False** Viruses cannot be artificially cultured in the laboratory. Furthermore, since they must be grown inside living tissue, it is difficult if not impossible to apply the postulates directly since tissues may harbor other latent viruses that were not detected. (p. 390; Figure 15.3)

3. **True** Some parasites use pili for attachment (*Neisseria gonorrhoeae*) and others use adhesive disks. The reaction allows them to obtain tissue fluids readily. (p. 398; Figure 15.8)

4. **False** Endotoxins are weaker than exotoxins; however, some such as found with organisms of typhoid fever and plague can easily kill in relatively large doses. (p. 394; Table 15.2)

5. **False** Cytopathic viruses are ones that cause observable changes in cells. They can be cytocidal when the virus kills the cell or noncytocidal if they do not. Noncytopathic would mean that the virus causes no observable changes. (p. 397; Figure 15.6)

6. **True** Herpes viruses often cause latent infections because they may not exhibit any symptoms for long periods of time prior to an outbreak. (p. 400)

7. **False** Although bacteria such as streptococci and staphylococci lyse blood cells, they do so in order to obtain iron which is necessary for their metabolism. (p. 395)

8. **d** Endotoxins commonly cause a rise in fever due to the toxic effect of the cell wall material. (p. 394; Table 15.2)

9. **e** Iatrogenic diseases are those that could be caused by poor medical treatment. If they are acquired during hospital treatment they are known as nosocomial infections. (p. 392)

10. **a** Some bacteria especially the *Clostridia* possess this spreading factor which enables them to digest hyaluronic acid. The acid appears in basement membranes of epithelial cells and acts to hold cells and tissues together. (p. 397)

11. **c** By subculturing organisms repeatedly on artificial media, the virulence of a pathogen can be greatly reduced. Also, the addition of chemicals such as formaldehyde can achieve the same result. (p. 388)

CHAPTER 16

EPIDEMIOLOGY AND NOSOCOMIAL INFECTIONS

One of the most important reasons that developed countries are as productive as they are today is that the population remains healthy and disease free. This essential task is performed by each country's health department and is carried out by individuals known as epidemiologists. Without their efforts and their coordination with others in the medical field, it would be very difficult, if not impossible, to obtain current information regarding important diseases, methods of transmission, and methods of control. Furthermore, information on the incidence or prevalence of diseases, and statistics on morbidity and mortality rates, all of which are essential to physicians and other medical personnel to help control and understand diseases, would not be available except through the efforts of the epidemiologists.

Prevention and control of hospital related or acquired infections is another important function of these specialists. These nosocomial infections can wreck havoc on already debilitated patients and must be controlled if there is any hope for successful treatment. This problem is vividly explained in the Microbiologist's Notebook on controlling infections in a burn unit.

The chapter then provides extremely valuable information relative to the role and activities of our public health departments and of the principles and techniques of disease control.

STUDY OUTLINE

I. **Epidemiology**
   A. What is epidemiology?
   B. Diseases in populations
   C. Reservoirs of infection
   D. Portals of entry
   E. Portals of exit
   F. Modes of transmission of diseases
   G. Disease cycles
   H. Public health organizations
   I. Notifiable diseases
   J. Epidemiological studies
   K. Control of communicable diseases
II. **Nosocomial Infections**
   A. General features
   B. Epidemiology of nosocomial infections
   C. Prevention and control of nosocomial infections

REVIEW NOTES

A. What is epidemiology, and what special terms are used by epidemiologists?

Epidemiology is the study of factors and mechanisms involved in the spread of disease within a population. Scientists that study epidemiology are epidemiologists and are concerned with the etiology or cause and transmission of diseases in the population. Incidence rates represent statistics relative to the number of new cases in a specific time period; prevalence rates reflect the number of people infected at any one time; morbidity rates indicate the number of cases as a proportion of the population, and mortality rates indicate the number of deaths as a proportion of the population. These statistics are reported in state health department reports and in the Morbidity and Mortality Weekly Report (MMWR) which is published by the Centers for Disease Control (CDC).

B. How are diseases categorized according to their spread in populations?

If a small number of isolated cases appear in a population, it is called a sporadic disease. If a large number of cases continually appear in a population but the harm to the patients is not too great, it is called an endemic disease. In epidemic diseases, a large number of cases suddenly appear in a population and patients are sufficiently harmed to create a public health problem. Common-source epidemics spread from a single source such as food or water, and propagated epidemics are spread by person-to-person contact. A pandemic is an epidemic disease that has achieved exceptionally wide geographic distribution.

C.   **How do various kinds of reservoirs of infection contribute to human disease?**

A reservoir is a site where organisms can persist and maintain their ability to infect. Reservoirs include humans, animals, and nonliving sources. In human reservoirs, carriers often transmit diseases passively, if they are unaware of the disease agent; actively, if they are recovering from the infection; or intermittently, if they release the pathogens periodically. Animal reservoirs can transmit diseases by direct contact or by means of insect vectors. These animal diseases are called zoonoses. See Table 16.1 for a list of selected zoonoses. Nonliving reservoirs include water, soil, or wastes.

D.   **What are the roles of portals of entry and exit and modes of transmission in the spread of human disease?**

The portals of entry include all those openings into the body such as the eye, ear, nose, and throat. They also include the skin, mucous membrane linings, and even the placenta. Although the latter protects the fetus, some diseases such as syphilis, toxoplasmosis, listeriosis, rubella, and AIDS can cross this barrier.

TABLE 16.2 Modes of transmission of selected diseases

Modes of Transmission	Examples of Diseses Transmitted
Contact transmission	
Direct contact	Rat-bite fever, rabies, syphilis, gonorrhea, herpes, staphylococcal infections, cutaneous anthrax, genital warts
Indirect contact by fomites	Hepatitis B, tetanus, rhinovirus and enterovirus infections
Droplets	Common cold, influenza, measles, Q fever, pneumonia, whooping cough
Vehicle transmission	
Foodborne	Intoxication with aflatoxins and botulinum toxin, paralytic shellfish poisoning, staphylococcal food poisoning, thyphoid fever, salmonellosis, listeriosis, toxoplasmosis, tapeworms
Waterborne	Cholera, shigellosis, leptospirosis, *Campylobacter* infections
Airborne, including dust particles	Chickenpox, tuberculosis, coccidioidomycosis, histoplasmosis, influenza, measles
Vector transmission	
Biological	Plague, malaria, yellow fever, typhus, Rocky Mountain spotted fever, Chagas' disease
Mechanical (on insect bodies)	*E. coli* diarrhea, salmonellosis, conjunctivitis

The portals of exit are generally the same as the portals of entry. Organisms usually exit in body fluids or feces.

Modes of transmission include direct contact, vehicle, or vector. **See Table 16.2** for the modes of transmission of selected diseases. Transmission by carriers, sexual practices, and transmission of zoonoses pose special epidemiologic problems.

E. **What is a disease cycle, and how is group immunity related to disease cycles?**

Many diseases such as chicken pox and influenza occur in cycles. In many cases, the cycle relates to the resistance status of the population. If many are not resistant and the agent returns, the disease can recur. If a large group within the population is resistant it can provide group or herd immunity and resist the agent even though some in the population are susceptible.

F. **How do the functions of organizations and the reporting of diseases contribute to public health?**

Public health organizations exist at city, county, state, federal, and world levels. They help to establish and maintain health standards, cooperate in the control of infectious diseases, collect and disseminate information, and assist with professional and public education. These organizations all contribute to the overall health and productivity of a community.

G. **What are the purposes and methods of epidemiological studies?**

Epidemiological studies help us to learn more about the spread of diseases in populations and how to control them. Descriptive studies note the number of cases of a disease, which segments of the population were affected, etc. Analytical studies focus on establishing cause and effect relationships in the occurrence of diseases in populations. Experimental studies are designed to test a hypothesis, often about the value of a particular treatment.

H. **What methods are used to control communicable diseases?**

Methods of control include isolation, quarantine, active immunization, and vector control. Most hospitals have infection control committees whose responsibility is to set guidelines and policies for isolation procedures of patients. See Table 16.4 for a summary of the important isolation procedures commonly used in hospitals. Quarantine is rarely used because of the availability of antibiotics. Active immunization is used to control many diseases including measles, mumps, rubella, and diphtheria. Vector control is effective where vectors can be identified and eradicated.

I.   What are nosocomial infections and how are they studied epidemiologically?

Nosocomial infections are acquired in a hospital or other medical facility. They can be exogenous, arising from bacteria outside the patient, or endogenous, arising from bacteria associated with the patient. Most are caused by <u>Escherichia coli</u>, <u>Staphylococcus aureus</u>, <u>Streptococci</u>, and <u>Pseudomonas</u>. Many strains are antibiotic resistant. Host susceptibility is an important factor in the development of such infections. **See Figure 16.21 below** for an illustration of the modes of transmission of nosocomial infections in a hospital setting and also refer to Figure 16.23 for a list of the most common sites of infections.

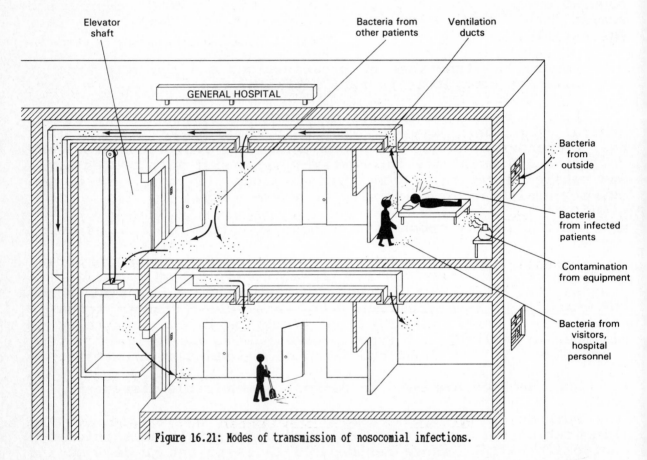

Figure 16.21: Modes of transmission of nosocomial infections.

J.   How can nosocomial infections be prevented and controlled?

Most hospitals have extensive infection control programs to ensure that proper hand washing, use of gloves, scrupulous attention to maintaining sanitary conditions and sterility where possible, and surveillance of antibiotic use and other hospital procedures are followed. By following good common sense, proper hand-washing procedures, and the universal precautions guidelines (see Table 16.5), most infections could be reduced dramatically.

154

LEARNING ACTIVITIES

Complete each of the following items by supplying the appropriate word or phrase.

1. The study of the factors and mechanisms involved in the spread of disease within a population is known as _Epidemiology_.

2. The causation of disease is known as _Ethiology_.

3. The number of new cases seen in a specific period of time is called the _Incidence_.

4. The number of people infected at any one time is the _Prevalence_ rate.

5. The number of cases in relation to the total number of people is the _Morbitity_ rate.

6. When a disease occurs at a constant rate in a particular geographic area it is called an _Endemic_.

7. A world-wide epidemic is known as a _Pandemic_.

8. Occurrence of isolated cases that do not appear to pose a threat to the population are known as _Sporadic_ cases.

9. An epidemic that arises from contaminated food is a _Common_ _Source_ epidemic.

10. Rodents that can carry and maintain the disease of plague act as _reservoirs_ of the disease.

11. A person who has contracted a disease and subsequently releases the organism after he has recovered is an _Active_ _Carrier_.

12. Diseases that can be transmitted from animals to humans are called _zoonoses_.

13. The sites at which organisms enter the body are the _Portal_ _of_ _Entry_.

14. Direct transmission occurring by handshaking or kissing is also called _Horizontal_ transmission, while diseases that spread from parent to offspring would be _Vertical_ transmission.

15. If a large proportion of the population is immune to a disease, the few susceptible individuals can be protected by _Herd_ or _group_ immunity.

155

16. Inanimate objects such as pencils, dishes, and bar soap act as _fomites_.

17. If diseases are transmitted as a result of inanimate objects, it would be _indirect_ transmission.

18. Living things, especially arthropods, can act to transmit disease as _vectors_.

19. A _biological_ _vector_ can actively transmit pathogens that use the vector as part of its life cycle.

20. A type of epidemiological study that lists the number of cases of a disease, the time period involved, and population segments affected is a _____ study.

21. The first case of a disease to be identified is often called the _____ case.

22. Epidemiological studies designed to test a hypothesis is an _____ study.

23. A nonmedicated substance given to a patient that has no effect on the patient but is perceived by the patient as a treatment would be called a _placibo_.

24. A patient with a highly infectious disease such as pneumonic plague or diphtheria must be placed under _strict_ isolation.

25. An individual who is separated from others due to a communicable disease is said to be in _quarantine_.

26. Hospital-acquired infections are known as _nosocomial_ infections.

27. Infections arising from the patient's own normal flora bacteria are _endogenous_ infections.

28. Special CDC guidelines that have been put into place to guard against the possibility of the AIDS virus being transmitted are known as _universal_ _precautions_.

Match the following terms with the description listed below. Choices may be used more than once.

Key Choices:

a. Prevalence
b. Propagated epidemic
c. Endemic
d. Sporadic
e. Mortality rate
f. Morbidity rate
g. Incidence rate
h. Pandemic
i. Common source epidemic
j. Zoonoses

_f_ 1. An increase of chickenpox from 6.7 cases/100,000 population to 15.4 cases/100,000 as reported in the weekly report from the CDC.

_b_ 2. A serious outbreak of rabies in skunks has led to an increase of the disease in domestic dogs.

_i_ 3. 67 people became sick with 4 deaths due to the consumption of botulism toxin in canned mushrooms.

_f_ 4. The disease of endemic typhus with a case rate of less than 1/100,000 in the northern half of the United States.

Match the following terms with the description listed below. Choices may be used more than once.

Key Choices:

a.	Direct contact	f.	Waterborne
b.	Indirect contact	g.	Airborne
c.	Fomites	h.	Foodborne
d.	Biological vector	i.	Droplets
e.	Mechanical vector		

_c,i_ 5. Streptococcal infections that occur in day-care centers.

_e,h_ 6. Flies that carry parasitic cysts and land on lettuce and other produce in an open-air market.

_a_ 7. Sexual transmission of herpes simplex.

_f_ 8. Dysentery resulting from improper chlorination of drinking water.

Match the following terms with the description listed below. Choices may be used more than once.

Key Choices:

a.	Strict isolation	d.	Enteric isolation
b.	Protective isolation	e.	Wound & skin isolation
c.	Respiratory isolation	f.	No isolation

_a,o,g_ 9. Requires hand washing.

_f_ 10. A noncommunicable disease such as tetanus.

_a_ 11. Requires a private room for adults.

_d_ 12. Necessary for cases of hepatitis and cholera.

157

___b___ 13. Essential for patients with burns, kidney failure, and with immune system defects.

___f___ 14. Required for patients with AIDS related complex (ARC).

REVIEW QUESTIONS

True/False (Mark T for True, F for False)

_____ 1. Fortunately, surgical masks, especially the cotton variety are capable of eliminating the aerosol spread of all microbes during surgery.

___F___ 2. WHO is a health organization that publishes a weekly newsletter called the Morbidity Mortality Weekly Report (MMWR). *Published by CDC.*

___T___ 3. Hand washing and the use of aseptic technique is required by all health personnel when working with patients placed in every type of isolation used in a hospital.

___T___ 4. Endogenous infections are often caused by opportunists from the patient's own normal flora.

___T___ 5. Generally speaking for disease agents, the portal of exit is usually the portal of entrance.

___F___ 6. The most common body site affected by nosocomial infectious agents is the gastrointestinal tract. *Urinary*

Multiple Choice

___c___ 7. Select the most common pair of nosocomial infectious agents.
   a. Escherichia coli, Proteus
   b. Staphylococcus aureus, Klebsiella
   c. Escherichia coli, Staphylococcus aureus
   d. Proteus, Pseudomonas

___d___ 8. In respect to nosocomial infections:
   a. catheterization rarely represents a site of infection because of the use of sterile disposable items.
   b. proper hand washing can ensure sterility and therefore eliminate this procedure as a source of infection.
   c. the most common site of nosocomial infections is the skin.
   d. the extensive use of antibiotics has greatly increased the chances of nosocomial infections.
   e. All of the above are false.

_e_ 9. Transmission of infectious diseases can occur via all the
following except:
a. indirect contact via droplets.
b. fomites.
c. mechanical vector transmission.
d. airborne via dust particles.
e. All the above are methods of transmission.

_a_ 10. A disease can be classified as a pandemic when:
a. it reaches worldwide distribution.
b. the mortality rate is very high.
c. it always occurs every year.
d. the disease affects the nervous system.

_c_ 11. Which of the following statements is not true concerning
universal precautions.
a. These precautions were the result of concerns about
the AIDS virus.
b. The precautions do not apply to feces, sputum, and
urine unless they contain visible evidence of blood.
c. The precautions do apply to blood, semen, vaginal
fluids, sweat, and tears regardless if visible blood
is present or not.
d. Universal precautions apply to all patients not just
those that are infected with AIDS.

ANSWER KEY

Fill-in-the-Blank

1. Epidemiology 2. Etiology 3. Incidence 4. Prevalence
5. Morbidity 6. Endemic 7. Pandemic 8. Sporadic
9. Common-source 10. Reservoir 11. Active carrier 12. Zoonoses
13. Portals of entry 14. Horizontal, vertical 15. Herd, group
16. Fomites 17. Indirect 18. Vectors 19. Biological vector
20. Descriptive 21. Index 22. Experimental 23. Placebo
24. Strict 25. Quarantine 26. Nosocomial 27. Endogenous
28. Universal precautions

Matching

1.f,b,g 2.b,j 3.i,a 4.d,f

5.a,c,i 6.e,h 7.a 8.f

9.a,b,c,d,e,f 10.f 11.a,b,c 12.d 13.b 14.f

Review Answers

1. **False** Surgical masks generally can impede the spread of microbes but will not prevent the spread of all droplets following a forceful sneeze. Cotton masks, since they can become moist, can reverberate and actually increase the spread. (p. 418; Figure 16.9)

2. **False** WHO stands for the World Health Organization. The MMWR is a weekly report published by the Centers for Disease Control in Atlanta, Georgia. (pp. 410, 423; Figure 16.1)

3. **True** Hand transmission is the most common method of transmission of infectious agents in the hospital. Proper hand washing and aseptic technique is mandatory for all types of patients regardless of whether they are under isolation or not. (pp. 429-430; Table 16.4)

4. **True** Endogenous organisms arise from one's own normal flora. Since we carry a number of infectious opportunistic agents such as S. aureus and E. coli, the likelihood of infection by these organisms once our resistance is lowered is very great. (p. 430)

5. **True** Organisms that gain access via the respiratory tract usually exit the same portal and ones affecting the gastro-intestinal tract exit via the oral/fecal route. (pp. 415-417; Figures 16.7, 16.8)

6. **False** The most common site of infection is actually the urinary tract followed by surgical wounds. (p. 433; Figure 16.23)

7. **c** Escherichia coli and Staphylococcus aureus are the most common agents because all of us carry these organisms as part of our normal flora (E. coli is in the intestinal tract and S. aureus is on our skin). They also are quite hardy and many strains are antibiotic resistant. (p. 430; Figure 16.20)

8. **d** The insertion of a catheter virtually insures an infection from organisms in the urethra. Hand washing does not insure sterility. The most common site of nosocomial infections is the urinary tract. (p. 432; Figure 16.23)

9. **e** Infectious diseases can be transmitted via contact, vehicle, and vector transmission methods. (p. 418; Table 16.2)

10. **a** Pandemics have occurred with diseases such as swine flu, cholera, malaria, and human plague with devastating results. These diseases affected dozens of countries and, in many cases, whole continents during major outbreaks. (p. 412; Figure 16.4)

11. **c** Although the precautions do apply to blood, semen, and vaginal fluids, they do not apply to sweat and tears unless visible blood is present. (p. 432; Table 16.5)

CHAPTER 17

HOST SYSTEMS AND NONSPECIFIC HOST DEFENSES

Considering the wide variety of substances microbes can produce that make us sick and considering the numerous avenues by which microbes invade our body, it would seem that we have little chance of survival. However, as will be described in this chapter, we have an equally effective variety of defense mechanisms which in almost all cases prevents or inhibits the invasive activities of these organisms. In reality, it is the exceptional microbe that can actually invade, establish, and cause disease.

Virtually every body system has elaborate defenses against invading microbes. Basically there are two lines of defenses. The first or nonspecific line, which is the subject of this chapter, consists of anatomical barriers, inhibitory body fluids, phago-cytosis, and inflammation. The second or specific line constitutes our immune response system and is the topic of the next chapter.

As this chapter will explain, as long as the body remains intact and healthy, our first line of defense should be all that is needed. However, it's sure nice to know that we still have a backup.

STUDY OUTLINE

I. **Nonspecific Versus Specific Host Defenses**
    A. General features
    B. Nonspecific defenses
    C. Specific defenses
II. **System Structure, Sites of Infection, and Nonspecific Defenses**
    A. Skin
    B. Eyes and ears
    C. Respiratory system
    D. Digestive system
    E. Cardiovascular system
    F. Nervous system
    G. Urogenital system
    H. Lymphatic system
III. **Phagocytosis**
    A. General features
    B. Kinds of phagocytic cells
    C. The process of phagocytosis
IV. **Inflammation**
    A. Definition and characteristics of inflammation
    B. The acute inflammatory process
    C. Repair and regeneration
    D. Chronic inflammation
V. **Fever**
    A. Normal body temperature
    B. Pyrogens
    C. Effects of fever
    D. Clinical approaches to fever
VI. **Molecular Defenses**
    A. Interferon
    B. Complement
    C. Acute phase response

REVIEW NOTES

A. **How do nonspecific and specific host defenses differ?**

The human body is protected by two basic defense systems. Nonspecific defenses operate regardless of the invading agent and constitute a first line of defense. These include the skin as an anatomical barrier, body fluids with antimicrobial substances, phagocytosis where a cell can engulf a foreign particle, and inflammation which accelerates the protective and healing processes. **See Figure 17.1** for an overview of the body's nonspecific defenses.

The second line of defense is a specific one which involves the immune defense system. It will be described in the next chapter.

162

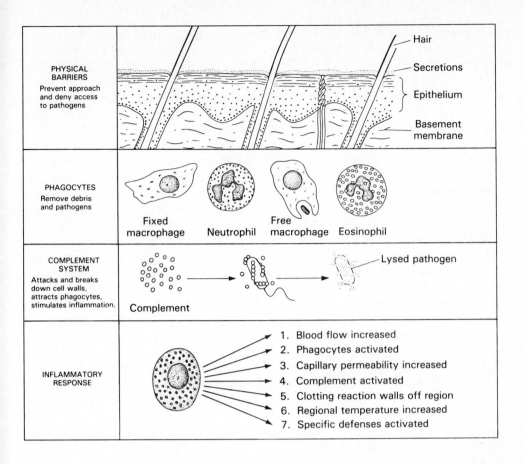

PHYSICAL BARRIERS Prevent approach and deny access to pathogens	Hair — Secretions — Epithelium — Basement membrane
PHAGOCYTES Remove debris and pathogens	Fixed macrophage Neutrophil Free macrophage Eosinophil
COMPLEMENT SYSTEM Attacks and breaks down cell walls, attracts phagocytes, stimulates inflammation.	Complement — Lysed pathogen
INFLAMMATORY RESPONSE	1. Blood flow increased 2. Phagocytes activated 3. Capillary permeability increased 4. Complement activated 5. Clotting reaction walls off region 6. Regional temperature increased 7. Specific defenses activated

Figure 17.1: An overview of the bodies nonspecific defenses.

B.   What are the important structures, sites of infection, and nonspecific defenses of the skin, eyes and ears, respiratory, digestive, cardiovascular, nervous, urogenital, and lymphatic systems?

The skin is well protected with a tough outer epidermis of keratin, acidic and salty secretions, and basement membranes between the outer layer and the dermis. Infections can occur anywhere the skin is broken, in ducts of glands, in hair follicles and sometimes on unbroken skin. See Figure 17.2 for an overview of the structure of skin and Figure 17.3 for a view of the sites of skin infections.

The eyes and ears are well protected with the eyes having eyelids, eyelashes, conjunctiva, a tough cornea, and lacrimal glands that secrete lysozyme enzymes. The ears are protected with hair and ceruminous glands in the ear canal. Eye infections can occur on the eyelids, conjunctiva, and cornea; ear infections can occur in the ear canal and middle ear. See Figures 17.4 and 17.6 for a view of the structure of the eye and ear.

The respiratory system is well protected by mucus and cilia, the mucociliary escalator, and phagocytic cells. Infections of the mucous membranes of the upper respiratory tract are common and can

163

spread to the sinuses, the middle ear, and to the lower respiratory tract. Infections here are often very severe especially to the elderly and to those with immune deficiencies. See Figure 17.7 for an overview of the respiratory system.

The digestive system is well protected with mucus, stomach acid, Kupffer cells in the liver, lymphatic tissue along the digestive tract, and by competition from normal flora bacteria. Infections can occur in the mucous membranes of virtually all organs. See Figure 17.8 for an overview of the digestive system.

The cardiovascular system is well protected by the cleansing effect of blood flow out of wounds, constriction of injured blood vessels, blood clotting, phagocytic action by certain classes of leukocytes, and by the release of heparin and histamine, which initiates the inflammatory response. The cardiovascular system is normally sterile, however, pathogens can be transported and can multiply in blood, infect the heart valves, and infect the pericardium. See Figure 17.10 for an overview of this system.

The nervous system is well protected with macrophages, the meninges, and the blood brain barrier. The system is normally sterile but pathogens can invade through the sinuses and blood and attack the meninges, the nerve endings, and even the brain tissue. See Figure 17.12 for an overview of this system.

The urogenital system is well protected by urinary sphincters, cleansing action of the outflow of urine, acidity of the urine and mucous membranes, and by competition from normal flora. Most infections seem to occur around the openings of the system. See Figure 17.13 for an overview of the urinary system.

The lymphatic system is well protected by the action of phagocytic cells. Since the lymphatic tissue filter the blood and lymph, they are susceptible to infection by pathogens that have gained access to the interiors of the body. See Figures 17.16 and 17.17 for overviews of the lymphatic system.

C. **What are the stages in the process of phagocytosis, and what kinds of cells are involved?**

Phagocytes are cells that ingest and digest foreign particles. Phagocytosis is the process by which this is accomplished. Phagocytic cells (macrophages and monocytes) are found fixed in the organs of the body, wandering in the tissues, and circulating in the blood stream.

The process of phagocytosis occurs when the invading microorganisms cause the release of chemical attractants called lymphokines which aid the macrophage in locating them (chemotaxis). The phagocyte then engulfs the particle and ingestion begins. A phagosome is formed. Lysosomes migrate to the phagosome and empty their digestive enzymes to begin the digestive process. Following digestion, the wastes are released by exocytosis. Microbes can resist phagocytosis by producing resistant capsules, preventing release of lysosomal enzymes, and by producing toxins to kill the phagocyte.

D.   What is inflammation?

Inflammation is a very complex process and is the bodies response to tissue damage, characterized by redness, swelling, heat, and pain. Inflammation can be either acute (short-term) or chronic (long-term).

E.   What are the steps in the acute inflammatory process and their functions?

Inflammation is initiated by some kind of cellular injury or death. When this occurs, histamine is released which initiates the events of inflammation. See Figure 17.21 for a summary of the steps in the process of inflammation and subsequent healing.

F.   How do repair and regeneration occur following acute inflammation?

The end result of inflammation is the repair and/or restoration of the damaged tissue. This occurs as capillaries grow into the site of injury and fibroblasts replace the dissolving blood clot. The resultant granulation tissue is strengthened by connective tissue fibers and the overgrowth of epithelial cells. See Figure 17.21 for a review of this healing process.

G.   How do the causes and effects of chronic and acute inflammation differ?

Acute inflammations such as bee stings, tend to occur very rapidly, reach a height very quickly, and subside quickly with very little tissue damage. On the other hand, chronic inflammations such as poison ivy tend to develop very slowly, reach a height very slowly, and take a long time to subside. It also exhibits a significant degree of tissue damage. Chronic inflammations occur when the host defenses fail to completely overcome the agent and may persist for years.

H.   How does fever function as a nonspecific defense?

Fever is an increase in body temperature caused by pyrogens increasing the setpoint of the temperature-regulating center in the hypothalamus. Exogenous pyrogens such as bacteria come from outside the body and endogenous pyrogens come from inside the body. Fever augments the immune response system and inhibits the growth of many microbes. It also increases the rate of chemical reactions and raises the temperature above the optimum growth rate for some pathogens. Antipyretics such as aspirin are recommended only for very high fevers and for patients with disorders that would adversely be affected by fever.

I.   How do interferon and complement function in nonspecific defenses?

Nonspecific defenses also include the molecules interferon and complement. Interferons are proteins produced by virus-infected cells that act nonspecifically to cause adjacent cells to produce antiviral proteins. Interferon is now made by recombinant DNA technology and may be very useful for certain therapeutic applications. See Figure 17.23 for an illustration of the mechanism by which interferon works. Notice that interferon itself does not inhibit viruses but its action on other cells causes the development of the resistance.

Complement refers to a series of serum proteins that when activated by an antigen/antibody reaction cause a sequence of events which includes increasing phagocytosis, inflammation, and cellular lysis. It is interesting to note that antibodies do not actually kill but only bind to antigen surfaces. Complement proteins are responsible for the cellular lysis. The complement system includes two pathways, the classic complement pathway and the properdin pathway. The action of the system is rapid and nonspecific. Deficiencies in any of the complement proteins can greatly reduce resistance to infection. See Figure 17.25 for a review of this system.

A third recently discovered nonspecific defense mechanism is acute phase response which involves several proteins including C-reactive protein (CRP). CRP seems to initiate an inflammatory response or accelerate an ongoing one. It also seems to activate the complement system, stimulate migration of phagocytes, and initiate platelet aggregation. CRP may also prevent death in certain otherwise fatal bacterial infections.

LEARNING ACTIVITIES

Complete each of the following items by supplying the appropriate word or phrase.

1. The ability of our body to defend against particular pathogens without having any particular experience with them is called _Resistance_.

2. An inhibitory substance produced by the oil glands is _Sebum_.

3. Another term for ear wax is _Cerumen_.

4. The inner ear is protected from external infection by the _tympanic_ _membrane_.

5. The respiratory system is lined with a protective _Serous_ membrane.

6. A mechanism which expels material out of the respiratory tract is the _mucociliary_ _system_.

7. A special coating of the digestive tract which can trap bacteria and prevent them from attaching to the surface is _____.

8. A type of leukocyte that possesses a granular cytoplasm and a multilobed nucleus is a _____ leukocyte.

9. The most numerous of all leukocytes which is also phagocytic is a _____.

10. A capillary barrier of the brain that limits entry of microbes and other substances is the _____ _____ _____.

11. The membranes surrounding the central nervous system that can be infected are the _____.

12. The urethra and bladder is generally protected from microbial invasion because of the _____ pH.

13. Peyers patches, tonsils, and the spleen are all part of the _____ system.

14. Lymphatic tissues contain wide passageways lined with phagocytic cells called _____.

15. The lymph tissue of the intestinal tract including Peyer's patches and other lymph nodes are part of the GALT or _____ _____ _____ _____.

16. A multilobed lymphatic organ, present at birth but atrophied by adulthood is the _____ gland.

17. A hormone produced by the thymus gland that stimulates production of lymphocytes in other tissues is _____.

18. The process of ingesting and digesting foreign particles by white blood cells is called __Phagocytosis__.

19. Monocytes that migrate into tissues to defend against microbes are _____ _____.

20. The macrophage system is known as the _____ _____.

21. WBC's that may be particularly important in parasitic worm infections are __Eosinophils__.

22. Macrophages can find microorganisms due to _____.

23. Chemical substances released by T-cells that act to combat foreign antigens are ___antibody___ .

24. Macrophages specifically found in the liver are _____ cells.

25. A vacuole formed within the cytoplasm following phagocytosis is a _____ .

26. A substance released by a bacterium that can kill a phagocyte is a ___leukocidins___ .

27. Redness, swelling, heat, and pain are signs of ___Inflammation___ .

28. A chemical released by injured cells that causes dilation of blood vessels is ___Histamine___ .

29. Pain associated with tissue injury may occur as a result of a kinin called _____ .

30. The ability of WBC's to pass through capillary walls is _____ .

31. An accumulation of pus in a cavity formed by tissue damage is an _____ .

32. Connective tissue cells that form fibrin material are called _____ .

33. A collection of necrotic tissue, phagocytes, and immune cells is a _____ .

34. Fever is caused by substances known as _____ .

35. A protein signal released by virus-infected cells is ___Interferon___ .

36. A series of regulatory proteins that bind in an antibody/antigen reaction are called ___Complement___ .

37. Antibodies that aid in phagocytosis are called _____ .

38. The two complement pathways are the _____ and the _____ pathways.

39. A laboratory test that involves the complement process is the ___Complement fixation___ test.

40. The acute phase response is characterized by a special protein called CRP or C - ___reactive protein___ .

168

Match the following terms with the description listed below.
Choices may be used more than once.

Key Choices:

a.	Skin	e.	Cardiovascular system
b.	Eyes and ears	f.	Nervous system
c.	Respiratory system	g.	Urogenital system
d.	Digestive system	h.	Lymphatic system

___f___ 1. Infections may be called meningitis and encephalitis.

___a___ 2. Acts as a mechanical barrier due to the presence of keratin proteins.

___h___ 3. System is normally sterile and devoid of microbes.

___d___ 4. System protected by acidic pH in part or whole.

___b___ 5. System is protected by lysozyme secretions produced by lacrimal glands.

___c___ 6. Opportunistic pathogens can be found on or around these systems.

Match the following terms with the description listed below.
Choices may be used more than once.

Key Choices:

a.	Phagocytosis	d.	Interferon
b.	Inflammation	e.	Complement
c.	Fever	f.	Pyrogens

___e___ 7. Involves opsonization.

___a___ 8. Neutrophils and monocytes can perform this function.

___d___ 9. Substance produced by virus-infected cells.

___b___ 10. Release of histamine causes this effect.

___f___ 11. Increases the rate of chemical reactions and activates the immune system.

REVIEW QUESTIONS

True/False (Mark T for True, F for False)

_____ 1. The sebum from oil glands creates a salty, alkaline environment which is inhibitory to microbes.

169

___I___ 2. Some pathogens can actually multiply within white blood cells.

acts indirectly c̄ infected cells

___F___ 3. Interferon seems to directly interfere with viral penetration of uninfected cells.

___I___ 4. Fever is now found to be beneficial in many infections and should not always be inhibited unless it goes above 40°C.

___I___ 5. Part of the "complement cascade" is designed to actually kill foreign cells by forming holes in cell membranes. *ASS*

in Phagocytosis

_____ 6. The eyes are protected by the enzyme amylase which is produced by the lacrimal gland and which causes lysis of cell membranes.

cellular eng

_____ 7. Acute inflammations are often characterized by the formation of granulomatous tissue followed by the release of histamine.

Multiple Choice

_____ 8. In the development of the immune system:
a. all animals and a few plants have been found to possess a specific immune system.
b. specific antibodies have been found in all vertebrates and in all types of fish.
c. Although all vertebrates have been found to be capable of rejecting grafts of foreign tissue, all invertebrates seem to lack this capacity.
d. Virtually all invertebrates appear to lack specific and nonspecific defenses but compensate for this by reproducing at extremely high rates.

_____ 9. Prostaglandins:
a. cause capillaries to dilate.
b. are released by virus-infected cells.
c. may intensify pain in an injured area.
d. are only produced by T-cells.

_____ 10. Which of the following is not a part of the effects of the complement proteins?
a. opsonization
b. inflammation
c. membrane lysis
d. interferon production
e. all of the above are major parts.

170

11. In terms of inflammation:
    a. the process is always detrimental to the host.
    b. aspirin should always be given especially for children to reduce the fever.
    c. inflammation should be suppressed to allow healing to occur.
    d. bradykinin injections should be given to alleviate the symptoms.
    e. all of the above are not true.

12. Which of the following is not a useful nonspecific defensive mechanism?
    a. Presence of a mucous lining of the respiratory system.
    b. Lysozyme present in tear fluid.
    c. Release of large quantities of histamine.
    d. Pepsin release in the stomach.

ANSWER KEY

Fill-in-the-Blank

1. Resistance 2. Sebum 3. Cerumen 4. Tympanic membrane
5. Serous 6. Mucociliary escalator 7. Mucin
8. Granular leukocytes 9. Neutrophil 10. Blood brain barrier
11. Meninges 12. Low 13. Lymphatic system 14. Sinuses
15. Gut-associated Lymphoid tissue 16. Thymus 17. Thymosin
18. Phagocytosis 19. Wandering macrophages
20. Reticulo-endothelial system 21. Eosinophils 22. Chemotaxis
23. Lymphokines 24. Kupffer cells 25. Phagosome 26. Leukocidins
27. Inflammation 28. Histamine 29. Bradykinin 30. Diapedesis
31. Abscess 32. Fibroblasts 33. Granuloma 34. Pyrogens
35. Interferon 36. Complement 37. Opsonins
38. Classical, properdin 39. Complement fixation
40. C-reactive protein

Matching

1.f 2.a 3.e,f,h 4.d,g 5.b 6.a,b,c,d,g

7.a,e 8.a 9.d 10.b 11.b,c,f

Review Answers

1. **False** Although sebum is inhibitory, the secretions are actually acidic (pH 3-5). The salt is produced by the sweat glands. (p. 446)

2.   **True** White blood cells can phagocytize foreign particles and bacteria.   Unfortunately some bacteria such as Rocky Mountain Spotted Fever rickettsia and the bacteria that cause tuberculosis can multiply within the white blood cells. (p. 458)

3.   **False** Interferon acts indirectly and is produced by virus-infected cells.   This protein is then absorbed by adjacent uninfected cells which stimulates them to produce antiviral proteins. (pp. 463-464; Figure 17.23)

4.   **True** Fever seems to increase the level of immune response and seems to inhibit the growth of microorganisms. (p. 462)

5.   **True** The complement cascade functions to assist phagocytosis, inflammation, and to initiate cell lysis.
(pp. 465-466; Figure 17.25)

6.   **False** The enzyme lysozyme is produced by the lacrimal gland which affects bacterial cell walls.   Amylase is produced by salivary glands and breaks down starch. (p. 446; Figure 17.4)

7.   **False** Acute inflammations are characterized by cellular damage followed by the release of histamine, however, the formation of granulomatous tissue or granulomas is more related to chronic inflammations. (pp. 459-461)

8.   **b**   Most invertebrates and plants appear to lack a specific immune system; most invertebrates can reject foreign grafted tissue; invertebrates do possess excellent nonspecific defense systems. (pp.467-468)

9.   **c**   Prostaglandins are cellular regulators which seem to intensify the effect of other pain-inducing substances called bradykinins. (p. 460; Figure 17.21)

10. **d**   Interferon is produced and released by a virus-infected cell. (pp. 463-464)

11. **e**   The inflammatory response is a response to an injurious agent and is designed to accelerate the immune response and phagocytosis.   Aspirin given to young children may cause Reye syndrome.   Suppression of the inflammatory response may result in a delay in the healing process. Bradykinins may increase the pain involved in the response. (pp. 459-462; Figure 17.21)

12. **c**   The release of histamine in large quantities would cause constriction of smooth muscle and dilation of capillaries.   This would lead to symptoms of asthma and anaphylactic shock, which would be detrimental. (pp. 445-450)

CHAPTER 18

IMMUNOLOGY I

BASIC PRINCIPLES OF SPECIFIC IMMUNITY AND IMMUNIZATION

One of the most complex and intriguing systems of the human body is the immune system. This intricate recognition and response system provides us with our best defense against almost all agents that have managed to bypass our first line of defense. Its importance is dramatically illustrated in cases of immune deficiencies such as occur with inherited disorders and with diseases such as AIDS and with its effect on preventing diseases through the use of vaccines.

Some of the most important and significant discoveries of this decade have occurred in this field. Furthermore, it is one of the most exciting and challenging fields of science today with hundreds of colleges, universities, and research laboratories uncovering important discoveries in all aspects of this discipline.

With this in mind, the chapter begins with a presentation of the fundamental concepts of immunity including a discussion of the kinds of immunity, characteristics of antigens and antibodies, and the features of antibody and cell-mediated immunity. The chapter also provides information relative to the concepts of immunization. Using this as a background, you will begin to understand how the immune system recognizes and responds to the invading foreign substances. This information will also provide the background needed to understand immune deficiencies and disorders that will be presented in the next chapter.

STUDY OUTLINE

I. **Kinds of Immunity**
   A. Innate immunity
   B. Acquired immunity
   C. Active and passive immunity
II. **Characteristics of the Immune System**
   A. Antigens and antibodies
   B. Cells and tissues of the immune system
   C. Dual nature of the immune system
   D. General properties of immune responses
III. **Humoral Immunity**
   A. General characteristics
   B. Properties of antibodies (immunoglobulins)
   C. Primary and secondary responses
   D. Kinds of antigen-antibody reactions
   E. Monoclonal antibodies
IV. **Cell-Mediated Immunity**
   A. General characteristics
   B. The cell-mediated immune reaction
   C. How killer cells kill
   D. Role of activated macrophages
   E. Comparison of function of B and T cells
   F. Factors that modify the immune response
V. **Immunization**
   A. Active immunization
   B. Passive immunization
   C. Future of immunization
VI. **Immunity to Various Kinds of Pathogens**
   A. Bacteria
   B. Viruses
   C. Fungi
   D. Protozoa and helminths

REVIEW NOTES

A. **What do immune, immunity, susceptibility, nonspecific immunity, specific immunity, immunology, and immune system mean?**

The word immune means free from burden. Immunity refers to the ability of an organism to recognize and defend itself against infectious agents. In other words, to be free of "burden" of these infectious agents. Susceptibility means to be vulnerable to disease agents. Nonspecific immunity is a defense against any infectious agent without having any experience with it, while specific immunity is a defense against specific agents generally because of some type of experience with the agent. Immunology is the study of specific immunity, and the immune system is the body system that provides the host with specific immunity to a particular infectious agent.

B.   How do innate immunity, acquired immunity, and active and
      passive immunity differ?

      Innate or genetic immunity exists because of genetically
determined characteristics.   In other words, we are simply born
with it and play no role in obtaining it.   We do not get diseases
that dogs or horses get simply because we are human.   Acquired
immunity is immunity obtained in some manner other than by heredity.   We acquire this after we are conceived or after we are born.
Active immunity means our body must make the antibodies while
passive means that we receive the preformed antibodies that were
made by someone else.   **Figure 18.1** illustrates the various types of
immunity and Table 18.1 lists the characteristics.

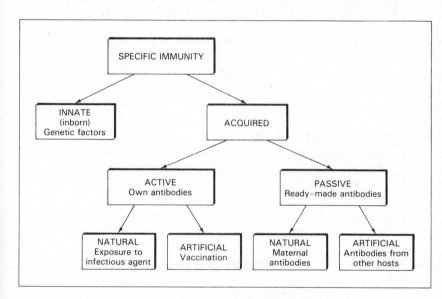

Figure 18.1: Types of immunities.

      Natural immunities are derived naturally such as actually
getting the disease while artificial immunity is usually obtained
by getting an injection of the material.

C.   What are the properties of antigens and antibodies, and how
      do cells and tissues function in the dual roles of the immune
      system?

      Antigens are substances that our immune system can recognize
and, if foreign, cause a specific immune response.   To be recognized as an antigen, it must be greater than 10,000 molecular
weight, it must have an electrical charge, and it must have
multiple antigenic determinants or epitopes.   The best antigens are
protein in nature but those that are made of glycoproteins, nucleo-
proteins, lipoproteins, or complex polysaccharide are quite good as
well.   Epitopes are found on large molecules such as bacterial
toxins, viruses, bacterial cells, animal cells, and, of course,
human cells.   See Figure 18.2.

175

Antibodies are proteins produced in response to the presence of a foreign antigen. Antibodies are capable of binding specifically to the epitopes found on the antigen surface.

Lymphocytes differentiate into B-cells in the bursal-equivalent (gut-associated lymphoid) tissues, or into T-cells in the thymus. Null cells remain undifferentiated.

The immune system consists of a dual or two-arm system. The humoral or antibody producing arm is carried out by B-cells, and the cell-mediated arm is carried out mainly by T-cells. See Figure 18.3 for an illustration of the differentiation of B and T cells. Table 18.5 provides a list of the characteristics of these cells.

D. How do recognition of self, specificity, heterogeneity, and memory function in the immune system?

Both the humoral and cell-mediated response systems based on the clonal selection theory have the ability to recognize self antigens from non-self antigens and to respond to them. However, the ability to respond to self antigens was probably destroyed during embryonic development by deleting those lymphocytes that were capable of responding to self antigens. Specificity refers to the ability of lymphocytes to respond to each antigen in a different and particular way. Heterogeneity refers to the ability of the immune system to produce many different substances such as antibodies in accordance with the many different antigens they encounter. The immune system also has the property of memory, that is, it can recognize substances it has previously encountered often at a faster rate. Table 18.3 has a summary of these attributes.

E. How do B cells and antibodies function in humoral immunity?

Humoral or antibody immunity represents one arm of the immune defense system. Upon sensitization of appropriate B-cells, specific antibodies are produced to the antigen that sensitized the B-cell. The appropriate B-cell is selected on the basis of the antibody present on the B-cell's membrane. This membrane antibody is somewhat like a sign board of a restaurant or bakery that indicates what it can do (in this case what type of antibody it can produce). When a B-cell is presented with an antigen it can react with, it binds with it, and divides many times to produce a clone of antibody-producing plasma cells and some memory cells. Helper T-cells are often required to assist in the binding of the antigen, while suppressor cells probably limit the duration of antibody production. The memory cells remain in the lymph tissue to respond to a later exposure to the same antigen. **See Figure 18.5** which provides an overview of this process.

Antibodies are Y-shaped protein molecules composed of four amino acid chains: two identical short or light chains and two identical long or heavy chains. The chains are held together by disulfide bonds. Figure 18.8 illustrates the structures of the different classes of antibodies, and Table 18.4 lists the properties of the molecules. Note that the IgG molecule represents the

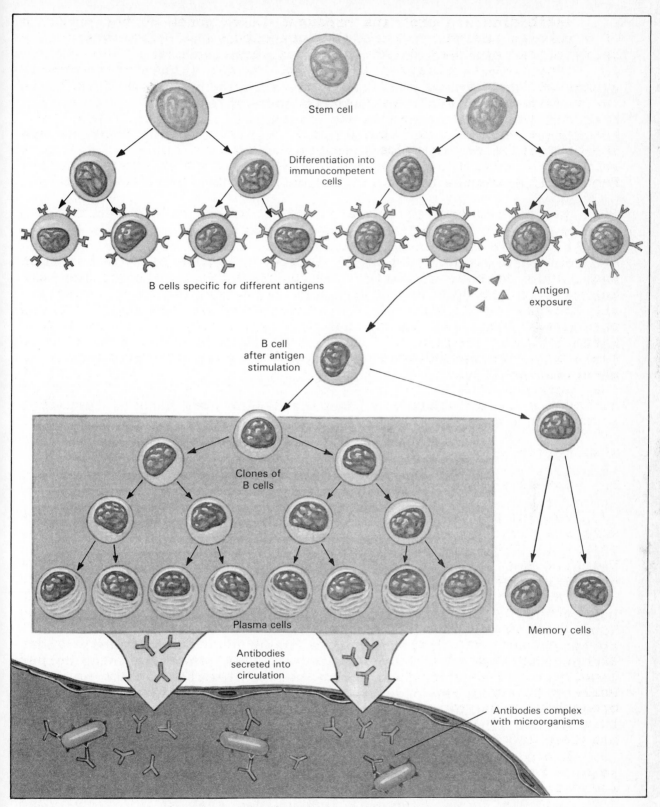

Figure 18.5: Clonal selection theory and B-cell response.

basic format. It's also the bodies primary antibody molecule.

The primary response is the response to the first exposure of the antigen. The secondary response is the increased response due to the presence of memory cells present as a result of the first exposure. See Figure 18.9 which summarizes these responses.

Antibody-mediated immunity is most effective against toxins produced by bacteria and acute bacterial and viral infections. Agglutination, lysis by complement, or neutralization reactions are most effective against these agents.

F. What are monoclonal antibodies, and how are they made and used?

Monoclonal antibodies are antibodies produced in the laboratory by hybrid cells called hybridomas. These cells contain the genetic information from both a myeloma (cancer) cell and a sensitized antibody-producing B-cell. The cancer cell provides the hybrid with the ability to produce large quantities of specific antibodies for virtually an infinite amount of time. These antibodies are used in diagnostic tests and for experimental methods in the treatment of certain diseases and cancers. Figure 18.14 illustrates the unique process used in the production of these antibodies.

G. How does cell-mediated immunity differ from humoral immunity, and how do reactions of cell-mediated immunity occur?

Cell-mediated immunity relates to the direct actions of sensitized T-cells that defend the body against chronic bacterial and virus infections, that reject tumors and transplanted tissues, and that cause delayed hypersensitivities. This immune response involves the differentiation and activation of several kinds of T-cells such as cytotoxic (T_c), delayed hypersensitivity (T_D), helper (T_H), and suppressor (T_s), and the secretion of lymphokines. T-cells do not make antibodies but do have membrane receptors for antigens and for histocompatibility proteins.

The reaction begins with the processing of an antigen by a macrophage, with the subsequent insertion of antigenic molecules into its own cell membrane. This antigen binds with T-cell receptors and with proteins on the membrane of macrophages. These macrophages then secrete interleukin-1 (IL-1) that activates helper T-cells. IL-1 and IL-2 (from helper t-cells) then activate cytotoxic T-cells, suppressor T-cells, and delayed hypersensitivity T-cells. IL-1, IL-2, and gamma interferon activate natural killer (NK) cells. The process is very complex, but the main idea is to specifically activate the proper type of T-cell so that the response to the foreign substance can be rapid and controlled. See Figure 18.15 which illustrates the action and interactions of several types of T-cells.

T_D cells secrete several lymphokines that stimulate various activities of macrophages. T_c cells can kill virus-infected cells and NK cells kill tumor cells. See Figure 18.17 for an illustration of the process of activation of various types of T, B, and natural killer cells. Note the chemical messengers or lymphokines that are used in each case.

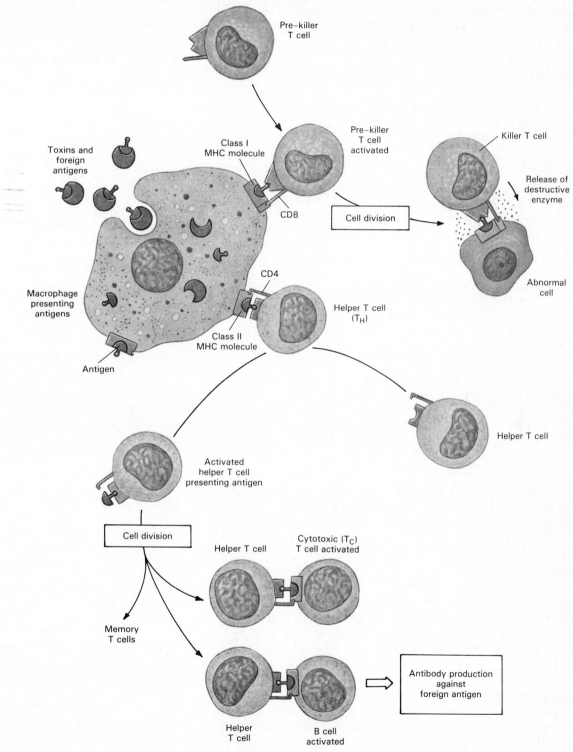

Figure 18.15: Cell mediated immunity.

H. **What are the special roles of killer cells and activated macrophages?**

Cytotoxic T (T_c) and natural killer (NK) cells destroy target cells by releasing the lethal protein perforin. Activated macrophages can specifically seek out a foreign substance such as a bacterium and phagocytize it. However, some pathogens can grow inside the macrophage, but an activating factor may help to stimulate antimicrobial processes so that the macrophage can kill the pathogens. Failure to do so, the macrophage may be walled off in the formulation of granulomas.

I. **What factors modify the immune responses?**

The nonspecific host defenses in healthy adults are usually sufficient to prevent nearly all infectious diseases. Individuals with reduced resistance are called compromised hosts. Factors that can reduce this are very young or old age, poor nutrition, lack of exercise, injury, pollution, and radiation. Immune system deficiencies such as a lack of certain complement proteins, infections such as AIDS, and genetic defects can also modify the response.

J. **What are the mechanisms of immunization, recommended immunizations, and the benefits and hazards of immunization?**

Active immunization occurs by the same mechanism as having a disease since it challenges the immune system to develop specific defenses and memory cells. However, it usually does not result in a permanent immunity with the first exposure and therefore requires subsequent "booster" shots. Vaccines made from live attenuated or dead organisms can be used as well as toxoids which are made from inactivated toxins. Table 18.8 lists the most commonly used vaccines in the United States. Note that they include the DPT and MMR series. The benefits of active immunization outweigh any of the hazards in their use. Vaccines made from cell wall material (whooping cough vaccine) are very difficult to purify and generally result in the greatest number of problems.

Passive immunization occurs by the same mechanism as natural passive transfer of antibodies in that the antibodies are pre-made. Passive immunity is provided by using gamma globulin shots, convalescent sera, or antitoxins made in other animals. Since the antibodies are pre-made, the protection is short-lived but is immediate. The main hazard comes from hypersensitivity.

Nonspecific and antibody mediated defenses are most important in controlling bacterial infections. Viral infections are also controlled by nonspecific and antibody mediated defenses and by interferon. Fungi, protozoa, and helminths are mostly controlled by cell-mediated defenses.

Efforts are being made to produce and purify vaccines that will not produce any unwanted side effects. These new preparations include subunit and recombinant vaccines which should be much safer than the currently used attenuated vaccines.

LEARNING ACTIVITIES

Complete each of the following items by supplying the appropriate word or phrase.

1. The ability of a host to recognize and respond to an infectious agent is called _Immunity_.

2. The study of the immune response is known as _immunology_.

3. An immunity common to all members of a species is _innate_ or _species_.

4. An immunity obtained by receiving preformed antibodies is called _passive_.

5. An immunity obtained by causing the host to produce antibodies is called _natural active_.

6. A substance that the body identifies as foreign is an _antigen_.

7. Antibody binding sites on the surface of antigens are called _Antigenic_ _determinant_ or _Epitopes_.

8. A molecule that is generally too small to qualify as an antigen may be called a _Hapten_.

9. The quantity of a substance needed to produce a reaction involving antibodies and antigens is measured as a _titer_.

10. Cells that are processed and matured in the bursal equivalent tissue are called _B cells_.

11. Cells that undergo differentiation in the thymus gland are called _T cells_.

12. Undifferentiated lymphocytes that are neither B nor T cells may be called _Null_ cells.

13. The two major types or "arms" of our immune system are the _antibody_ and _cell-mediated_ immunities.

14. Normal host tissues and substances can be identified by our immune response system as _self_.

15. The most plausible theory to explain how the body can recognize self from non-self is the _Clonal Selection_ theory.

16. The ability to remove those cells programmed to destroy host tissues is known as ___Tolerance___.

17. The ability of the immune system to recognize and respond quickly to a foreign substance due to a previous encounter is called ___Memory___.

18. B-cells that actually produce antibodies are ___Plasma___ cells.

19. The H and L chains of antibody molecules are held together by ___Secretory___ bonds.

20. The component of IgA molecules that facilitates transport across cell borders is the ___secretory___ component.

21. B-cells generally require a ___T helper___ T-cell to aid in production of antibodies.

22. An antibody-antigen reaction involving large antigens such as bacteria or blood cells results in a visible response called ___Agglutination___.

23. Specific antibodies produced by a single clone of cultured cells are ___Monoclonal___.

24. An antibody-antigen reaction involving bacterial toxins can usually be detected by a test called a _____.

25. A lymphokine that appears to activate helper T-cells is called _____.

26. T-cells that are capable of a direct response against foreign cells by killing them are called ___Cytotoxic___ T-cells.

27. Killer cells contain granules of a lethal protein called _____.

28. A substance that contains an antigen to which the immune system responds is a ___Vaccine___.

29. In the DPT vaccine, the P stands for ___Pertussis___ which is the disease of ___Whooping___ ___Cough___.

30. In the MMR vaccine, the R stands for ___Rubella___ which is the disease of ___German___ ___measles___.

31. A vaccine used to protect against tuberculosis is the ___BCG___.

32. Antibodies made against specific toxins are called ___Antitoxins___.

33. A vaccine prepared by neutralizing the toxin is a _Toxoids_.

34. Two recently developed types of vaccine preparations which should reduce unwanted side effects are _____ and _____ vaccines.

35. An _antiserum_ is a serum that contains antibodies toward a particular antigen.

Match the following terms with the description listed below. Choices may be used more than once.

Key Choices:

    a. Innate immunity   d. Natural active acquired
    b. Acquired immunity   e. Artificial active acquired
    c. Natural passive   f. Artificial passive acquired
       acquired

b, c   1.   Antibodies received by an infant during breast feeding.

b, f   2.   An immunity obtained with a gamma globulin shot.

b, d   3.   An immunity by actually getting chickenpox.

a   4.   Human immunity to canine distemper.

b, e   5.   Getting the DPT shot.

Match the following terms with the description listed below. Choices may be used more than once.

Key Choices:

    a. IgA   d. IgD
    b. IgM   e. IgE
    c. IgG

c   6.   Crosses the placental barrier to provide antibody protection to the fetus.

d, a, e   7.   Possesses 2H and 2L chains.

b   8.   The first antibody secreted to fight infectious agents.

e   9.   Binds to mast cells and basophils and involved in allergies.

d   10.   Is rarely secreted and is found mainly on B-cell membranes.

b   11.   Held together by J chains.

_a_ 12. Found in tears and saliva.

Match the following terms with the description listed below. Choices may be used more than once.

Key Choices:

a. Helper T-cell
b. Suppressor T-cell
c. Delayed hypersensitivity
   T-cell
d. Memory T-cell
e. Natural killer cell

f. Cytotoxic T-cell
g. Plasma cell
h. Memory B-cell
i. Macrophage

_e, f, a_ 13. Can specifically kill infected host cells.

_g_ 14. Produces highly specific antibodies.

_h, d_ 15. Is formed after recognition and response to a foreign antigen.

_a, b, e, c_ 16. Secretes lymphokines.

_i_ 17. Can engulf foreign antigens and destroy them.

_h_ 18. Possesses surface receptors for IgG.

_g, h_ 19. Possesses antibody on their cell membranes.

_a_ 20. Attacked by the HIV in particular.

_a_ 21. Secrete interleukin 2.

REVIEW QUESTIONS

True/False (Mark T for True, F for False)

_F_ 1. Acquired immunities obtained by natural passive means generally last a lifetime.

_T_ 2. The upper ends of the Y of an antibody contain the combining sites.

_T_ 3. Although B and T cells are greatly involved in the immune response, macrophages and other phagocytes are essential to act upon and destroy bacteria and virus-infected cells.

_T_ 4. The MMR series contains live attenuated viruses.

___F___ 5. In most cases, the antibody molecules will actually kill the invading microbe immediately upon contact.

___F___ 6. The BCG vaccine for tuberculosis is now being used in the United States and is given to school children in inner city schools due to the increase in tuberculosis.

Multiple Choice

___C___ 7. Which of the following is not located on the Fc portion of the IgG molecule?
a. Opsonization site
b. Complement binding site
c. Secretory transport unit
d. Allergic reaction site
e. All of the above are found on the Fc portion.

___A___ 8. An important vaccine that is given to young children to prevent meningitis is:

a. Hib	c. DPT
b. MMR	d. BCG

_____ 9. In human lymphoid tissue, B cells are most abundant in the __?__ tissue while T cells are most abundant in the __?__ tissue.

a. Thymus/Peyer's patchs	c. Lymph nodes/spleen
b. Spleen/blood	d. Peyer's patchs/thymus

_____ 10. Which of the following is not a good characteristic of an antigen.
a. protein composition
b. molecular weight of 1500
c. multiple epitopes
d. having charged groups such as amine and carboxyl groups
e. Actually all of the above are good characteristics.

___C___ 11. IgG is characterized by:
a. having four combining sites.
b. having a G-shape to the protein chains.
c. having 2H and 2L chains.
d. acting as the secretory antibody.
e. all of the above are not true.

___C___ 12. T_D cells can perform all the following except:
a. stimulates phagocytes by producing MAF.
b. acts in a delayed-type hypersensitive response.
c. produces specific antibodies against foreign cells.
d. produces lymphokines to help macrophages find microbes.

_d_ 13. Vaccines made of live viruses:
   a. are always the best choice to develop immunity.
   b. always provide permanent immunity.
   c. do not cause any adverse symptoms to recipients.
   d. can be dangerous to pregnant women.

_a_ 14. In the cell mediated immune reaction:
   a. the foreign antigens are processed and presented on the cell membrane surface of a macrophage.
   b. the foreign antigen is first engulfed by a null cell prior to processing by a macrophage.
   c. when T cells bind to macrophages they quickly divide and transform into plasma cells.
   d. macrophages lack histocompatibility proteins which allow them to bind with any foreign substance.

ANSWER KEY

Fill-in-the-Blank

1. Immunity 2. Immunology 3. Innate, species 4. Passive
5. Active 6. Antigen 7. Antigenic determinants, epitopes
8. Haptin 9. Titer 10. B-cells 11. T-cells 12. Null cells
13. Antibody, cell-mediated 14. Self 15. Clonal selection theory
16. Tolerance 17. Memory 18. Plasma 19. Disulfide
20. Secretory component 21. Helper 22. Agglutination
23. Monoclonal 24. Neutralization 25. Interleukin-1
26. Cytotoxic 27. Perforin 28. Vaccine
29. Pertussis, whooping cough 30. Rubella, German measles 31. BCG
32. Antitoxins 33. Toxoid 34. Subunit, recombinant 35. Antiserum

Matching

1.b,c 2.b,f 3.b,d 4.a 5.b,e

6.c 7.a,c,d,e 8.b 9.e 10.d 11.a,b 12.a

13.e,f,i 14.g 15.d,h 16.a,b,c,e,f 17.i 18.e,f,i 19.g,h 20.a 21.a

Review Answers

1.   **False** Immunities of this type last only a few days or weeks. The antibodies are pre-made, that is, made by someone else. Therefore the body is not stimulated to develop a response to the foreign agent. (p. 474; Table 18.1)

2.   **True** The upper ends of all antibody molecules consist of variable regions of the heavy and light chains and differ from antibody to antibody. Therefore they can act as the combining sites. (pp. 481-484; Figure 18.7)

3.   **True** Macrophages are generally the first cells to respond to a foreign antigen and act to process it. The cells than seek out appropriate T or B cells for a cellular or antibody response. Once a response is made, macrophages also become involved to specifically engulf and destroy the invading agent. (pp. 481, 489)

4.   **True** The MMR vaccine is for measles, mumps, and rubella and is given routinely to children after 15 months of age. (p. 495; Table 18.8)

5.   **False** Antibodies bind to antigen but do not actually "kill" the foreign substance. Macrophages, killer T-cells, or the complement cascade are responsible for the "killing" effect. (pp. 481, 484)

6.   **False** The BCG vaccine is not licensed for use in the United States. If it were used, all recipients would skin test positive for tuberculosis. (p. 497)

7.   **c**   Secretory transport units are associated with IgA dimers not with IgG units. (p. 484)

8.   **a**   MMR is used for measles, mumps, and rubella; DPT is used for diphtheria, pertussis, and tetanus; and BCG is used for tuberculosis. (p. 496)

9.   **d**   While T cells and B cells are found in all these tissues, B cells are most abundant in Peyer's patches (60%) and T cells are most abundant in the thymus (99%). (Table 18.2, p. 478)

10. **b**   Antigens, to be recognized by the immune system must have molecular weights greater than 10,000. (p. 475)

11. **c**   IgG possesses two combining sites, a Y-shaped structure, and acts as the late responder to antigenic stimulation. (pp. 481-484; Figure 18.7, Table 18.4)

12. **c** T_D cells do not produce antibodies but do produce a great variety of lymphokines. Plasma cells produce specific antibodies. (pp. 489-490)

13. **d**   Live viruses are very useful for immunization purposes but should not be given to pregnant women because of the possibility that the virus could cross the placenta and infect the fetus. (pp. 497-498)

14. **a**   Foreign antigens are first processed by macrophages; when T cells bind with macrophages they divide and differentiate into other T cells such as cytotoxic and memory T cells; macrophages have histocompatibility proteins which allows them to bind to the appropriate T cell receptors. (Figure 18.15, pp. 489-492)

CHAPTER 19

IMMUNOLOGY II

IMMUNOLOGIC DISORDERS AND TESTS

Although the immune system is most noted for its beneficial effects, unfortunately it also makes us painfully aware of its detrimental effects. Immunological disorders including hypersensitivities, autoimmunities, and immunodeficiencies all have become important problems that require considerable amount of research and effort in their treatment. However, many of these disorders have yet to be fully understood.

Sometimes the immune response system can be suppressed to provide us opportunities for organ and tissue transplants. Without the significant advances and discoveries made in this area, heart, liver, and lung transplants would still be only ideas rather than realities.

The immune system can also be used effectively for identification purposes. The last part of this chapter presents the basic concepts of serology and outlines the most significant tests employed in hospital laboratories for the identification of a great variety of diseases.

In summary, this chapter presents both the challenges associated with disorders of the immune system and the successes that have been achieved in identifying and correcting them.

STUDY OUTLINE

I. **Overview of Immunologic Disorders**
    A. Hypersensitivities
    B. Immunodeficiencies

II. **Types of Hypersensitivities**
    A. Immediate (Type I) hypersensitivity
    B. Cytotoxic (Type II) hypersensitivity
    C. Immune complex (Type III) hypersensitivity
    D. Cell-mediated (Type IV) hypersensitivity

III. **Autoimmune Disorders**
    A. General features
    B. Spectrum of disorders
    C. Autoimmunization
    D. Examples of disorders

IV. **Transplantation**
    A. General features
    B. Histocompatibility antigens
    C. Transplant rejection
    D. Immunosuppression

V. **Drug Reactions**
    A. General features
    B. Types of reactions

VI. **Immunodeficiency Diseases**
    A. General features
    B. Deficiencies of cells of the immune system
    C. Acquired immunodeficiencies

VII. **Immunologic Tests**
    A. Serology
    B. Precipitin reactions
    C. Agglutination reactions
    D. Other reactions

REVIEW NOTES

A. What are the different types of immunologic disorders?

An immunologic disorder results from an inappropriate or an inadequate immune response. Most disorders involve either hypersensitivities which are inappropriate reactions to an antigen or allergen, and immunodeficiencies which are inadequate immune responses. **Table 19.5** provides an excellent summary of the four basic types of hypersensitivities. Types I and IV are generally the most common.

Immunodeficiencies can be primary in which the patient lacks T-cells or B-cells or has defective ones due to a genetic or developmental defect or secondary in which the patient has defective T- or B-cells after they have developed normally.

TABLE 19.5 Characteristics of the types of hypersensitivity

	Type I	Type II	Type III	Type IV
Characteristic	Immediate	Cytotoxic	Immune Complex	Delayed
Main mediators	IgE	IgG, IgM	IgG, IgM	T cells
Other mediators	Mast cells, anaphylactic factors, eosinophils	Complement	Complement, inflammatory factors, eosinophils, neutrophils	Lymphokines, macrophages
Antigen	Soluble or particulate	On cell surfaces	Soluble or particulate	On cell surfaces
Reaction time	Seconds to 30 minutes	Variable, usually hours	3 to 8 hours	24 to 48 hours
Nature of reaction	Local flare and wheal, airway constriction, anaphylactic shock	Clumping of erythrocytes, cell destruction	Acute inflammation effects	Cell-mediated cell destruction
Therapy	Desensitization, antihistamines, steroids	Steroids	Steroids	Steroids

B.   **What are the causes, mechanisms, and effects of immediate hypersensitivity?**

Immediate (Type I) hypersensitivity (anaphylaxis in its worst form) is the result of an inappropriate immune response to a harmless substance called an allergen. **Figure 19.1 below** summarizes the mechanism of this type of hypersensitivity. Note that the patient must have an initial exposure or sensitization which causes the release of IgE antibodies. The second exposure provokes the release of histamine and other chemicals which mediates the response.

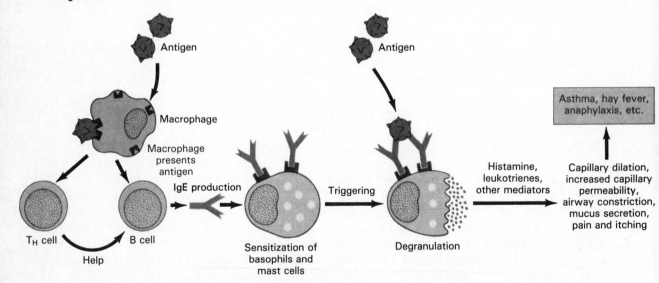

Figure 19.1: Mechanism of Type I hypersensitivity

Atopy is a localized reaction to an allergen such as pollen in which histamine and other mediators elicit sneezing and congestion of sinuses and other signs typical of this allergy. A skin reaction would exhibit a wheal and flare pattern.

Generalized anaphylaxis is a life-threatening systemic reaction in which blood pressure is greatly decreased or the airway is occluded.

Treatment requires desensitization as shown in Figure 19.5. Symptoms can be relieved with antihistamines or epinephrine. Table 19.5 summarizes the characteristics of Type I hypersensitivity.

C. **What are the causes, mechanism, and effects of cytotoxic reactions?**

Cytotoxic or Type II hypersensitivity involves specific antibodies reacting with cell-surface antigens detected as foreign by the immune system. The mechanism is summarized in Figure 19.6. Transfusion reactions and hemolytic diseases of the newborn are examples of this type of hypersensitivity.

D. **What are the causes, mechanism, and effects of immune complex disorders?**

Immune complex (Type III) hypersensitivity results from the formation of antigen-antibody complexes. The mechanism of this disorder is summarized in Figure 19.9. Serum sickness and the Arthus reaction which is illustrated in Figure 19.10 and 19.11 are examples of this type of hypersensitivity.

E. **What are the causes, mechanism, and effects of cell-mediated reactions?**

Cell-mediated (Type IV) is also known as delayed hypersensitivity because the reactions take more than 12 hours to develop. This hypersensitivity is mediated by T_D cells rather than IgE antibodies. See Table 19.5 which summarizes the basic characteristics of this hypersensitivity. The mechanism of cell-mediated hypersensitivity is summarized in **Figure 19.12 below**. Note that a sensitized T_D cell releases lymphokines which in turn activate macrophages to mediate the response.

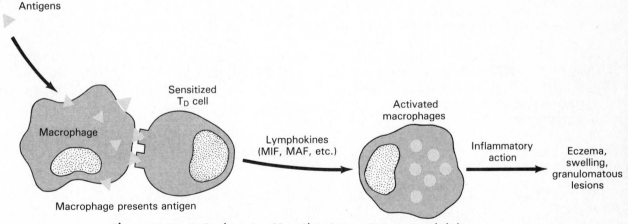

Figure 19.12: Mechanism of cell-mediated Type IV hypersensitivity.

Contact dermatitis such as seen with poison ivy, metals, rubber material, etc., are common examples. Others include hypersensitivity to tuberculin material, leprosy, fungal agents, and granulomatous hypersensitivity.

F.   How does autoimmunization occur, and how is hypersensitivity involved with it?

Autoimmunization is the development of hypersensitivity to self; it occurs when the immune system responds to a body component as if it were foreign. Normally all those cells capable of responding to self antigens are destroyed prenatally.

Tissue damage from these autoimmune disorders can by caused by cytotoxic, immune complex, and cell-mediated hypersensitivity reactions. Examples include myasthenia gravis; rheumatoid arthritis, which is illustrated in Figure 19.15; and systemic lupus erythematosus. The mechanism for the development of autoimmunities is not clear and may be different in different types of conditions.

G.   Why are organ transplants sometimes rejected, and how can rejection be prevented?

The ability to recognize self from non-self is the fundamental basis of the immune system. Genetically determined histocompatibility antigens (HLA's) produced by genes of the major histocompatibility complex (MHC) are found on the surface membranes of all cells and can be recognized by immune surveillance cells. These antigens in graft tissue are the main cause of transplant rejection, but immunocompetent cells in bone marrow grafts sometimes can destroy host tissue, resulting in a dangerous condition called graft vs. host.

Transplant rejection can be controlled by immunosuppression, which is a lowering of the responsiveness of the immune system to materials it recognizes as foreign. This is accomplished by radiation and by cytotoxic drugs such as cyclosporin. The use of these techniques can also reduce the host's immune response to infectious agents.

H.   How is hypersensitivity involved in drug reactions?

Most drug molecules are too small to act as allergens. However, if a drug combines with cellular proteins, it can be large enough to be recognized and a response can occur. All four types of hypersensitivity reactions have been observed in immunologic drug reactions.

I.   What are the causes, mechanisms, and effects of immunodeficiency diseases?

Immunodeficiency diseases can arise from defects in any part of the immune system and include deficiencies in T-cells, B-cells, complement proteins, macrophages, and others. T-cell deficiencies lead to a lack of cell-mediated immunity; B-cell deficiencies lead to a lack of antibody-mediated immunity; complement deficiencies lead to problems in opsonization, inflammation, and cellular destruction of pathogens.

Immunodeficiencies can be hereditary or acquired as a result of infections, malignancies, and autoimmune disorders. See Figure 19.20 which illustrates the various kinds of immunodeficiencies.

J.   How can antigens and antibodies be detected and measured?

Serology is the use of laboratory tests to detect antigens and/or antibodies. Detection is possible because of the high degree of specificity of the antigen/antibody reaction and because the combination of these molecules forms a lattice network that can be observed. Diagnosis can be made on the basis of seroconversion.

Precipitation reactions involve tiny soluble antigens that precipitate from solutions or in agar gels. They must reach an appropriate equilibrium to be visible as shown in Figure 19.23. Examples include immunodiffusion, immunoelectrophoresis and radial immunodiffusion.

Agglutination reactions involve large insoluble antigens such as blood cells or bacteria that can form large clumping, visible particles.

Complement fixation tests involve the use of complement proteins and blood cells to indirectly determine the presence of certain antibodies.

Immunofluorescence shown in Figure 19.29, allows detection of products of immune reactions within tissues or with cells by means of fluorescent dyes.

Other special tests include radioimmunoassay which uses radioactivity and enzyme-linked immunosorbant assay tests which use the high degree of specificity of enzymes in the detection of antigen/antibody reactions.

K.   What is AIDS and why is it causing an epidemic?

Most AIDS cases are caused by human immunodeficiency viruses (HIV-1 and HIV-2) which have evolved to attack the immune system. It is an infectious disease which gradually destroys the victims T_4 or helper cell population. This leaves the patient susceptible to infections caused by opportunistic agents. The agent is transmitted by contact with infected blood and body fluids and can also be acquired by a fetus carried by an HIV positive woman. Because it can be easily spread via blood transfusions, sexual contact, infected needles, etc., it has reached epidemic proportions in many countries.

L.   How can AIDS by diagnosed, treated, and prevented?

AIDS infections are detected by immunologic tests such as the ELISA test for the presence of antibodies. The disease is treated currently with AZT, ddC, and interferon but cannot be cured.

Prevention is to avoid exposure to infected body fluids. A vaccine has not yet been developed.

LEARNING ACTIVITIES

Complete each of the following items by supplying the appropriate word or phrase.

1. An exaggerated response to an antigen that the immune system normally ignores is a _Hypersensitivity_.

2. When a person has a defect in the B or T cell population it is called an _immunodeficiency_.

3. Another term for immediate hypersensitivity is _Anaphylaxis_.

4. The antibodies responsible for type I hypersensitivity were first known as _Reagins (IgE)_

5. An ordinary foreign substance that elicits an adverse immune response in a sensitized person is a _Allergen_.

6. The initial recognition of an allergen that promotes IgE production is _Sensitization_.

7. When mast cells release histamine, the process is known as _____.

8. Localized, "out of place", allergies are referred to as _Atopy_.

9. The current method of treatment for allergies involves _____.

10. The treatment for allergies involves the formation of antibodies called _blocking_ antibodies.

11. A reaction that occurs when matching antigens and antibodies are present in the blood at the same time is called _____ reaction.

12. Type III hypersensitivity results in the formation of large antibody-antigen complexes called _immune_ complexes.

13. A hyperimmune reaction to horse protein used in treatment of diseases such as diphtheria and tetanus is called _Serum_ _Sickness_.

14. A positive skin test toward a sample of horse serum injected under the skin would be seen as a _____ and _____.

194

15. Cell-mediated immunity is also known as ___delayed___ because it may require 2-3 days to elicit a response.

16. In a positive tuberculin skin test, a raised hard, red region of the skin is an ___Induration___.

17. Antibodies directed against self antigens are called ___Autoantibodies___.

18. An autoimmune disease that affects the skeletal muscle is ___Rheumatoid___ ___Arthritis___.

19. An autoimmune disease that affects the joints is _____ _____.

20. Grafts made between genetically identical individuals are called _____.

21. The use of pigskin to cover a burned area on a patient would involve a graft called a _____.

22. When cells from a tissue graft react negatively against the host it is called ___Graft___ ___ ___host___.

23. Human histocompatibility antigens identified specifically on leukocytes are called _____ _____ antigens.

24. The genetic complex that encodes the genes for histocompatibility is known as the _____ _____ _____.

25. In order to minimize transplant rejections in recipients, a process of ___Immuno Suppression___ is carried out.

26. A drug that suppresses the immune system, specifically the T-cell population is _____ ___.

27. Abnormalities in the embryonic development of the host such as failure of the thymus gland to develop are _____ immunodeficiencies.

28. The absence of B-cells and therefore antibodies in a patient is called _____.

29. A branch of immunology that involves laboratory testing of serum samples is ___Serology___.

30. An antibody/antigen reaction that involves the formation of tiny lattice-like networks of molecules is a ___Precipitation___ reaction.

31. A process of separating antigen molecules in a gel using an electric current and then reacting them with antibody is called _____.

32. An antibody/antigen reaction involving the agglutination of red blood cells is _____.

33. An antibody/antigen reaction which indirectly detects antibodies by determining whether complement added to the test is used up in the reaction is the _____ _____ test.

34. A neutralization test involving diphtheria toxin injected under the skin is the _____ test.

35. The use of fluorescent tagged antibodies or antigens is the basis of _____.

36. The use of an enzyme as an indicator after an antibody/antigen reaction has taken place is the basis of the _____ test.

37. A type of cancer common to AIDS patients is _____ _____.

Complete each of the following abbreviations by supplying the appropriate words.

38. HLA: _____.

39. MHC: _____.

40. AIDS: *Acquired Immune deficiency Syndrome*.

41. ELISA: *Enzyme linked immunoabsorbent Assay*.

42. GVH: _____.

43. SLE: _____.

44. SRS-A: _____.

45. Rh: _____.

46. MCTD: _____.

47. RA: _____.

48. SCID: _____.

49. RIA: _____.

50. HIV: _____.

51. ARC: _____.

52. AZT: _____.

53. PCP: _____.

Match the following terms with the description listed below.

Key Choices:

a. Immediate (Type I)   d. Delayed (Type IV)
b. Cytotoxic (Type II)   e. Autoimmune disorder
c. Immune complex (Type III) f. Immunodeficiency
                              g. GVH disease

__f__ 1.   Agammaglobulinemia

__e__ 2.   Rheumatoid arthritis

__c__ 3.   Serum sickness

__d__ 4.   Anaphylaxis

__g__ 5.   Rejection of the recipient by the transplant tissue.

__b__ 6.   Hemolytic disease of the newborn.

__a__ 7.   Circulation of sensitizing IgE antibodies.

__d__ 8.   Contact dermatitis and tuberculin hypersensitivity

__e__ 9.   Passive transfer through blood transfusions.

Match the following terms with the description listed below.
Choices may be used more than once.

Key Choices:

a. Precipitation test   e. Hemagglutination
b. Agglutination test   f. Complement fixation
c. Neutralization   g. Immunodiffusion
d. ELISA test   h. RIA test

_____ 10. Schick test for diphtheria.

_____ 11. Blood typing.

_____ 12. Separation of antigens by means of an electric current
           followed by antibody reaction.

197

_____ 13. Use of an enzyme as an indicator of the antibody/antigen reaction.

_____ 14. Measurement of antibody by means of radioactivity.

_____ 15. Zone of equivalence, visible as a hazy ring in a capillary tube.

_____ 16. The use of sheep red blood cells and complement.

REVIEW QUESTIONS

True/False (Mark T for True, F for False)

__T__ 1. Atopic allergies are often genetically based.

__F__ 2. Immunodeficiency diseases are always inherited.

__T__ 3. A major difference between hay fever and a common cold is that hay fever exhibits greater numbers of eosinophils in nasal secretions.

__F__ 4. A useful treatment for individuals with severe allergies to insect stings are syringes filled with epinephrine and SRS-A.

__T__ 5. HLA antigens are determined by a set of genes designated A,B,C,D and DR.

__F__ 6. A granulomatous response is usually associated with Type I hypersensitivities.

__F__ 7. An elevation of the CD4 cell count can be used to predict the onset of disease symptoms of AIDS.

Multiple Choice

__b__ 8. Women who receive Rhogam shots are attempting to control:
   a. severe asthma attacks.
   b. hemolytic disease of newborn.
   c. serum sickness.
   d. Arthus reactions.

__d__ 9. All the following relate to Type IV hypersensitivities except:
   a. responses elicited in 2-3 days.
   b. granulomatous responses.
   c. mediators include lymphokines and macrophages.
   d. passive transfer of circulating reagins.
   e. All of the above relate to Type IV.

198

_d_ 10. Autoimmunization:
  a. occurs when we vaccinate ourselves.
  b. involves GVH responses.
  c. is a process of hypersensitivity to pollen.
  d. is the basis for rheumatoid arthritis and systemic
     lupus erythematosus.

_d_ 11. All the following are true concerning immunosuppression
        during organ transplants except:
  a. may require radiation of the recipient to destroy the
     immune system.
  b. may leave the patient susceptible to infections.
  c. may result in the transplant tissue rejecting the
     recipient.
  d. may result in an autoimmune disease called myasthenia
     gravis.

ANSWER KEY

Fill-in-the-Blank

1. Hypersensitivity 2. Immunodeficiency 3. Anaphylaxis
4. Reagins 5. Allergen 6. Sensitization 7. Degranulation
8. Atopy 9. Desensitization 10. Blocking 11. Transfusion
12. Immune 13. Serum sickness 14. Wheal & flare 15. Delayed
16. Induration 17. Autoantibodies 18. Myasthenia gravis
19. Rheumatoid arthritis 20. Isografts 21. Xenografts
22. Graft vs host 23. Human leukocyte antigens 24. MHC
25. Immunosuppression 26. Cyclosporine A 27. Primary
28. Agammaglobulinemia 29. Serology 30. Precipitation
31. Immunoelectrophoresis 32. Hemagglutination
33. Complement fixation 34. Schick 35. Immunofluorescence
36. ELISA 37. Kaposi's sarcoma

38. Human leukocyte antigen
39. Major histocompatibility complex
40. Acquired immune deficiency syndrome
41. Enzyme linked immunosorbant assay
42. Graft versus host
43. Systemic lupus erythematosus
44. Slow releasing substance - A
45. Rhesis
46. Mixed connective tissue disease
47. Rhematoid arthritis
48. Severe combined immunodeficiency disease
49. Radioimmunoassay
50. Human immunodeficiency virus
51. AIDS related complex
52. 3'azido-3'thymidine
53. Pneumocystis carinii pneumonia

199

Matching

1.f 2.e 3.c 4.a 5.g 6.b 7.a 8.d 9.a

10.c 11.e,b 12.a,g 13.d 14.h 15.a 16.f

Review Answers

1.   **True**   Atopy means "out of place" and is often genetically related. At least 60% of those with atopy have a family history of the disorder. (pp. 512-513)

2.   **False** Immunodeficiency diseases may be acquired as a result of infection, malignancies, autoimmune diseases, or other conditions. (pp. 530-531)

3.   **True** An elevated eosinophil count can also suggest allergy or infection with eukaryotic parasites such as trichinella worms. (p. 513)

4.   **False** Severe allergic attacks can be treated with antihistamine or epinephrine but not SRS-A since this substance is one of the mediators in the allergic response. (pp. 514-515)

5.   **True**   HLA refers to human leukocyte antigens which are determined by a set of genes designed A,B,C, and D. Some of the genes are highly variable and may specify a variety of antigens. (p. 526; Figure 19.17)

6.   **False** Granulomatous hypersensitivities are the most serious of the cell-mediated Type IV hypersensitivities. (p. 521)

7.   **False** AIDS viruses specifically damage T4 (helper) cells because they bear a CD4 receptor. A drop in this cell count indicates the onset of disease symptoms. (pp. 539-540)

8.   **b**   Rhogam shots are attempting to control the Type II, cytotoxic reactions seen when an Rh-negative mother carries an Rh-positive fetus. (p. 516; Figure 19.2)

9.   **d**   Type IV hypersensitivities can be transferred through bone marrow transplants or by transfer factor. Type I hypersensitivities can be passively transferred. (p. 522; Table 19.5)

10. **d**   Autoimmunization refers to a process by which hypersensitivity to self-antigens develops. It is not clear whether it is the cause or the result of some other factors. (p. 522; Table 19.6)

11. **d**   Immunosuppression is the process of minimizing an immune response to a foreign substance. Generally, immunosuppression does not involve autoimmune diseases. (pp. 527-528)

CHAPTER 20

DISEASES OF THE SKIN AND EYES; WOUNDS AND BITES

The first line of defense against microbial invaders is our skin and mucous membranes. As long as this tissue is clean, healthy, and untraumatized, it is an effective barrier to almost all types of invaders. As will be presented in this chapter a few microbes, especially the invasive parasites, have found methods to attack and penetrate this virtual fortress either independently or by means of assistance as with arthropod vectors. For the most part, however, some type of damage to this tissue such as a cut or burn is necessary before the majority of bacteria and viruses are able to cause injury. When they do, the results are often very severe.

This chapter will then present a survey of the most important microbes that are able to attack and penetrate the defenses of the skin and eyes and cause serious diseases. Following this presentation, a short section will be provided describing those agents capable of infecting wounds or invading the tissues following arthropod bites.

The complete eradication of any disease is understandably difficult. However, the essay at the end of this chapter dramatically points out that diligent efforts on the part of governments, epidemiologists, and the populace as a whole against disease agents, in this case smallpox, can achieve this ultimate goal.

STUDY OUTLINE

I.   **Diseases of the Skin**
   A.   Bacterial skin diseases
      1.   Staphylococcal infections
      2.   Streptococcal infections
   B.   Viral skin diseases
      1.   Rubella
      2.   Measles
      3.   Chicken pox and shingles
      4.   Other pox diseases
      5.   Warts
   C.   Fungal skin diseases
      1.   Dermatophytes
      2.   Subcutaneous infections
      3.   Other fungal skin diseases
II. **Diseases of the Eyes**
   A.   Bacterial eye diseases
      1.   Ophthalmia neonatorum
      2.   Bacterial conjunctivitis
      3.   Trachoma
   B.   Viral eye diseases
   C.   Parasitic eye diseases
III. **Wounds and Bites**
   A.   Wound infections
   B.   Arthropod bites and infections

REVIEW NOTES

A.   **What kinds of pathogens cause skin diseases?**

Various bacterial, viral, fungal, and parasitic organisms cause skin diseases. Table 20.2 provides a complete summary of the diseases, the causative agents, and their characteristics.

B.   **What are the important epidemiologic and clinical aspects of skin diseases?**

Bacterial skin diseases are usually transmitted by direct contact, droplets, or fomites. Many of the agents are part of the normal flora of the skin that have gained access via cuts or scratches. Most of these diseases can be treated with penicillin or other antibiotics. Some bacteria such as Staphylococcus have become very resistant to many antibiotics because of their inappropriate use. Burn infections often are nosocomial and are commonly caused by antibiotic-resistant organisms such as Pseudomonas and Staphylococcus.

Viral skin diseases such as rubella, measles, and chicken pox are usually transmitted via nasal secretions. Some, such as small pox and warts are commonly transmitted by direct contact. Treatment of these skin diseases is difficult and usually confined to

202

relieving the symptoms and not curing the disease. Warts can be excised but commonly recur.

Fungal skin diseases are commonly acquired by direct contact and by fomites. Some fungal agents such as <u>Candida</u> are opportunists. Healthy, intact skin and mucous membranes are usually adequate to resist most of these agents. Topical antifungal agents are usually effective in controlling the dermatophytes such as athlete's foot and ringworm. Subcutaneous fungal infections often persist in spite of treatment with topical fungicides. Systemic infections such as blastomycosis and systemic candidiasis are very difficult to treat and often require Amphotericin B.

Table 20.2 provides additional epidemiological characteristics of these disease agents.

C.   **What kinds of pathogens cause eye diseases?**

Diseases of the eyes are caused by bacteria, viruses, and parasites. Table 20.3 provides a complete summary of the most important agents, the disease they cause, and their characteristics.

D.   **What are the important epidemiologic and clinical aspects of eye diseases?**

Bacterial eye diseases can be transmitted by direct contact, fomites, insect vectors, and to infants during delivery. Most are treated with antibiotics and prevented by common sense and good sanitation.

Viral eye diseases are transmitted by dust particles or direct contact. Since viruses do not respond to antibiotics, there is no real effective treatment available, but good common sense and proper sanitation can help to prevent these diseases.

Parasitic eye diseases are usually transmitted by insect vectors. By avoiding the insect bites and by controlling them, the incidence of these diseases can be reduced.

Table 20.3 provides additional information concerning each of these disease agents.

E.   **What kinds of pathogens infect wounds and bites?**

Wound infections can be caused by various types of bacteria such as <u>Clostridium</u>. Bite infections can be caused by ticks, chiggers, mites, fleas, and others. Often the bite wounds can be susceptible to further bacterial infection. Table 20.4 provides a complete summary of the wound infectious agents and the arthropods that cause bite wounds and infections.

F.   **What are the important epidemiologic and clinical aspects of wound and bite infections?**

Wound infections such as gas gangrene and other anaerobic infections can be caused by various species of soil bacteria

203

entering uncleaned lacerated wounds. Proper cleaning and draining of these wounds followed by treatment with antibiotics such as penicillin or antitoxins can reduce or eliminate the possibility of severe complications. Cat scratch and rat bite fevers can be prevented by avoiding the animals that can cause the injuries and by proper treatment of the affected areas. Laboratory workers who must work with these animals should be aware of the dangers of these infections and should take all necessary precautions including the use of protective clothing.

Arthropod bites and infections can be prevented by good sanitation and proper hygiene along with protection of the skin to avoid the bites.

Table 20.4 provides additional information relative to these agents.

LEARNING ACTIVITIES

Complete each of the following items by supplying the appropriate word or phrase.

1. An infection at the base of an eyelash is commonly called a
    _____.

2. Blackheads that become infected by the bacterium _Propionibacterium_ can result in _____.

3. The toxin responsible for scarlet fever is called the
    _____ toxin.

4. A highly infectious pyoderma caused by a staphylococcus or streptococcus is __Impetigo__.

5. The removal of dead tissue following a severe burn is called
    __Debridement__.

6. Rubella is also known as __German__ measles.

7. Red measles viruses can cause two severe complications, measles __encyphalitis__ and _____
    _____ _____.

8. Two to three days before the onset of symptoms of the red measles, red spots with bluish central specks called __Koplick__ spots, appear on the oral mucosa.

9. An adult case of chicken pox often results in painful lesions called __Shingles__.

10. Human papillomaviruses cause skin eruptions called __Warts__.

11. Tinea pedis is also known as __Athlete's__ __Foot__.

12. Fungi that invade keratinized tissue are called _Dermatophytes_ .

13. An oral infection caused by <u>Candida</u> <u>albicans</u> is _Thrush_ .

14. A foot infection caused by fungi of the genus <u>Madurella</u> is _Madura_ foot.

15. Schistosome larvae that normally infect birds can cause a skin infection in humans called _Swimmer's_ _itch_ .

16. An inflammation of the cornea of the eye is called _____ .

17. Bacterial conjunctivitis which causes the eye to become red and inflamed is commonly known as _Trachoma_ .

18. A serious eye infection typified by a "pebbled" or rough appearance is _____ .

19. Onchocerciasis caused by a nematode larvae is known as _____ _____ .

20. Deep wound infections caused by species of <u>Clostridia</u> that result in production of gas are _gas_ _gangrene_ .

21. Tissues filled with gas bubbles that snap, crackle, and pop are called _Clepitant_ tissues.

22. Spirillar fever is also known as _____ in Japan.

23. Scabies is caused by a _____ .

24. An infection caused by fly larvae is _____ .

25. "Nit-picking" means to remove _____ .

26. A louse infestation is called _____ .

27. Epidemic keratoconjunctivitis, sometimes called "ship yard eye", is most likely caused by a _____ .

Match the following terms with the description listed below. Choices may be used more than once.

Key Choices:

a.	Staphylococcus aureus	h.	Sporothrix shenckii
b.	Streptococcus pyogenes	i.	Candida albicans
c.	Pseudomonas aeruginosa	j.	Aspergillus
d.	Rubeola	k.	Schistosomes
e.	Rubella	l.	Dracunculus medinensis
f.	Varicella zoster	m.	Trichophyton
g.	Poxviridae		

_b_ 1.   Bacteria that can cause erysipelas.

_d_ 2.   Microbes that can cause a skin rash.

b, a 3.   Bacteria that can cause impetigo.

_i_ 4.   The organism that causes a yeast infection.

_e_ 5.   The German measles virus that can cause congenital defects.

_c_ 6.   Causes a greenish discoloration of tissue with a grape-like odor.

_j_ 7.   Causes a fungal infection that can ulcerate the eardrum.

_h_ 8.   Contracted from sphagnum moss; invades lymphatics.

_a_ 9.   Causes scalded skin syndrome.

_k_ 10. The larvae causes swimmer's itch.

e, g, d 11. Controlled by means of a live attenuated viral vaccine.

_l_ 12. These worms can be extracted through the skin by winding on a stick.

_b_ 13. Causes scarlet fever.

_m_ 14. Causes athletes foot.

_f_ 15. Causes chicken pox.

Match the following terms with the description listed below.

Key Choices:

a. Bacterium
b. Virus
c. Fungus
d. Chlamydia

e. Arthropod
f. Soil actinomycete
g. Helminth

_b_ 16. Red measles

_b_ 17. Warts

_a_ 18. Acne

_c_ 19. Barber's itch

_a_ 20. Boil

_g_ 21. Swimmer's itch

_f_ 22. Madura foot

_d_ 23. Trachoma

_b_ 24. Shingles

_c_ 25. Ringworm

_c_ 26. Thrush

_b_ 27. Smallpox

_a_ 28. Pinkeye

_e_ 29. Chiggers

_g_ 30. Loa Loa

_e_ 31. Scabies

_a_ 32. Gas gangrene

REVIEW QUESTIONS

True/False (Mark T for True, F for False)

_F_ 1. Once a person develops scarlet fever due to a strepto-coccal infection, he will never get a streptococcal infection again.

_T_ 2. Following severe burns, it is essential to remove the eschar to avoid serious bacterial infections.

_T_ 3. Chicken pox and shingles are actually caused by the same virus.

_F_ 4. Ringworm is actually caused by a fungus. *Dermatophytes*

_F_ 5. The genus name for the most common fungal agents that affect the skin is <u>Tinea</u>.

_F_ 6. Ophthalmia neonatorum is an eye infection caused by a species of <u>Chlamydia</u> similar to the one that causes trachoma.

Multiple Choice

___c___ 7. Molluscum contagiosum:
   a. causes warts.
   b. affects soft-bodied animals with hard shells.
   c. is a member of the pox virus family.
   d. can be treated with antibiotics.

___c___ 8. Pygmies of Uganda attained their short status by:
   a. being vegetarians.
   b. being infected with a tapeworm.
   c. having a growth hormone deficiency due to a nematode infection.
   d. having an addiction to hallucinogenic drugs.

___d___ 9. Dermatophycoses:
   a. are all caused by the same fungus.
   b. often have a high mortality rate.
   c. treatment usually requires amphotericin B.
   d. can cause infection in laboratory workers who are culturing the infectious agents.
   e. none of the above are true.

___b___ 10. The virus that causes warts:
   a. can only be transmitted sexually.
   b. can sometimes spontaneously disappear.
   c. can be treated with acyclovir.
   d. can only invade the respiratory tract.

___e___ 11. An appropriate treatment for chickenpox in young children could be:
   a. tetracyclines   c. amantadine
   b. aspirin   d. ivermectin
                             e. none of above

___e___ 12. Which of the following would represent the least likely way of contracting an eye infection by a pseudomonad or other opportunistic pathogen.
   a. frequent use of a hot tub
   b. wearing mascara
   c. wearing contact lenses
   d. using testers at cosmetic counters
   e. All of the above are good ways to contract eye infections.

CLINICAL PERSPECTIVE

The following clinical case was based on an actual case presented in the Morbidity Mortality Weekly Reports (MMWR) published by the Centers for Disease Control (CDC). Some of the details have been changed to better illustrate the type of data that may be obtained from the patient. This case should illustrate how epidemiological, clinical, cultural, and other types of information can be used to develop a complete diagnosis and treatment of a disease.

Following the presentation of the case, a series of questions will be asked relative to the diagnosis, suggested laboratory tests and types of therapy received. Try to answer these yourself before reading the actual answers. At the end of the case, a short, follow-up analysis and additional comments will be provided to help correlate all the information about the disease.

CASE HISTORY

Two fraternity brothers reported to the health clinic at their college in Ohio with lymphadenopathy, low-grade fever, conjunctivitis, sore throat, and mild arthralgia. One patient exhibited a slight maculopapular rash on the face and scalp. They could not remember the date of their measles, mumps, rubella (MMR) vaccinations and did not receive any recent influenza vaccine.

Preliminary diagnosis?

Influenza, red measles (rubeolla), German measles (rubella).

Additional case information:

The fraternity brothers reported that a student from South America had visited with them about a week before they experienced any symptoms. They did not recall the visitor nor anyone else being sick or experiencing any other similar symptoms.

Suggested laboratory tests?

Serological hemagglutination-inhibition (HI) antibody and ELISA tests for influenza, rubeolla, and rubella.

Laboratory report:

Positive rubella specific IgM titer with a greater than fourfold rise in HI antibody titer.

Probable diagnosis?

German measles

Suggested treatment?

Supportive therapy and bed rest unless there are serious complications.

Follow-up and analysis:

Both students recovered without complications. They probably contracted the disease from the visiting student as no other fraternity member reported similar symptoms. Proof of rubella immunity was provided by only 32.5% of the fraternity members even though the school had a rubella immunization requirement for entry.

Free vaccine was made available as MMR or measles rubella (MR) by the health service clinic to prevent an outbreak but only 45 students out of 35,000 accepted. About 65% of U.S. colleges now have requirements for measles and/or rubella immunity; however, most have no enforcement nor use the immune status as a condition of enrollment.

ANSWER KEY

Fill-in-the-Blank

1. Sty 2. Acne 3. Erythrogenic 4. Impetigo 5. Debridement
6. German 7. Encephalitis, subacute sclerosing panencephalitis
8. Kopliks 9. Shingles 10. Warts 11. Athletes foot
12. Dermatophytes 13. Thrush 14. Madura 15. Swimmer's itch
16. Keratitis 17. Pinkeye 18. Trachoma 19. River blindness
20. Gas gangrene 21. Crepitant 22. Sodoku 23. Mite 24. Myiasis
25. Lice 26. Pediculosis 27. Adenovirus

Matching

1.b 2.d,e,g,f 3.a,b 4.i 5.e 6.c 7.j 8.h 9.a 10.k 11.d,e,g 12.l 13.b
14.m 15.f

16.b 17.b 18.a 19.c 20.a 21.g 22.f 23.d 24.b 25.c 26.c 27.b 28.a
29.e 30.g 31.e 32.a

Review Answers

1. **False** A person will not get another case of scarlet fever due to antibodies produced against the toxin. However, because there is a great variety of strains of streptococci, anyone acquiring a streptococcal infection will be just as likely to get another one. (p. 551)

2. **True** The scab or eschar that forms over a severe burn can represent a site for infection beneath the tissue. Pathogens could then gain access to the blood. (p. 552)

3.   **True**   The varicella-zoster virus causes both diseases. Chicken pox usually occurs in children and zoster in adults. (p. 554)

4.   **True**   Ringworm is not caused by a worm but by a fungus such as Epidermophyton, Microsporum, and Trichophyton. (p. 558; Table 20.1)

5.   **False**   Tinea is actually a synonym for cutaneous infections. It is used along with another term to identify a cutaneous fungal infection on some portion of the body. (p. 558)

6.   **False**   Ophthalmia neonatorum is caused by a pyogenic diplo-coccus, Neisseria gonorrhoeae. (pp. 561-562)

7.   **c**   Human papilloma viruses cause warts.   Molluscum contagiosm cannot be treated by antibiotics nor any other treatment. (pp. 556-557)

8.   **c**   When mothers are infected with the nematode, Onchocerca volvulus, the parasite damages the pituitary gland of their fetuses thus causing a deficiency of the growth hormone. (p. 563)

9.   **d**   Dermatophycoses is caused by a great variety of fungal agents; rarely cause any deaths except by secondary infection; and can be treated with topical, antifungal agents. (p. 557; Table 20.1)

10.   **b**   The wart virus can be transmitted by direct contact or by fomites; warts can be treated with cryotherapy but may reappear; and they can grow freely in several body areas. (pp. 556-557)

11.   **e**   Tetracyclines could cause discoloration of developing teeth; aspirin could cause Reye syndrome in young children; amantadine is useful for influenza A infections; ivermectin is an antihelminthic drug. (p. 555)

12.   **e**   Improperly chlorinated hot tubs, contaminated mascara, contaminated and improperly cleaned contact lenses, and even testers at cosmetic counters could easily contribute to eye infections. (pp. 564-565)

CHAPTER 21

DISEASES OF THE RESPIRATORY SYSTEM

Among the most common diseases to affect humans are colds. It seems we never can avoid these pesky viruses no matter what we do. In addition, if our resistance is lowered due to stress, poor health, or lack of sleep, our chances of "catching" these viruses is greatly increased. Perhaps this is Mother Nature's way of trying to slow our hectic pace before more serious types of illnesses can "catch" up with us.

In this chapter, not only will there be a description of the important agents that cause upper respiratory infections, but also all those agents that affect the lower respiratory tract will be described. Those diseases of highest incidence or greatest importance will be stressed; however, even those diseases that are rather uncommon, or perhaps occur elsewhere in the world, will be presented to ensure a thorough survey of all those agents capable of affecting the respiratory system.

The essay at the end of this chapter concerning the influenza pandemic of 1918 provides a reminder that deadly illnesses of past decades should never be forgotten.

STUDY OUTLINE

I. **Diseases of the Upper Respiratory Tract**
   A. Bacterial upper respiratory diseases
      1. Pharyngitis and related infections
      2. Diphtheria
      3. Ear infections
   B. Viral upper respiratory diseases
      1. The common cold
      2. Parainfluenza

II. **Diseases of the Lower Respiratory Tract**
   A. Bacterial lower respiratory diseases
      1. Whooping cough
      2. Classic pneumonia
      3. Mycoplasma pneumonia
      4. Legionnaires' disease
      5. Tuberculosis
      6. Ornithosis
      7. Q fever
      8. Nocardiosis
   B. Viral lower respiratory diseases
      1. Influenza
      2. Other viral infections
   C. Fungal respiratory diseases
      1. Coccidioidomycosis
      2. Histoplasmosis
      3. Cryptococcosis
      4. Pneumocystic pneumonia
   D. Parasitic respiratory diseases

REVIEW NOTES

A. **What bacteria cause upper respiratory infections and what are the important epidemiologic and clinical aspects of these diseases?**

Various species of bacteria cause upper respiratory infections and most of these such as species of <u>Streptococcus</u>, <u>Mycoplasma</u>, <u>Staphylococcus</u>, and <u>Haemophilus</u> are somewhat common inhabitants. Infections related to pharyngitis include laryngitis, epiglottitis, sinusitis, and bronchitis. Table 21.1 provides a complete summary of the agents, the diseases they cause, and the characteristics of the disease.

These infections are exceedingly common and are easily acquired through inhalation of droplets, especially in winter when people are crowded indoors. Most of these infections can be treated with penicillin or some other broad-spectrum antibiotic; however, complete eradication of the infectious agent or agents is rare. Much depends on the condition of the respiratory system and the host's immune defense system.

Diphtheria is no longer common in the United States because of

the use of the diphtheria toxoid (the "D" of DPT shot). Both the
organism and its toxin which is produced by a lysogenic prophage,
contribute to the symptoms. This includes the formation of a
grayish white pseudomembrane on the pharynx and massive swelling
leading to a symptom called "bullneck." The disease is spread by
droplets and can be treated with antitoxin and antibiotics.

Ear infections occur in the middle and outer ear and are very
common in young children. Organisms reach the middle ear via the
eustachian tube and cause severe pain and swelling. Most infections respond to penicillin or other antibiotic. In severe cases,
artificial openings or "tubes" (see Figure 21.4) can be made in the
eardrum to relieve the pain and fluid buildup.

**B.   What viruses cause upper respiratory infections and what are
the important epidemiologic and clinical aspects of these
diseases?**

Numerous viral groups affect the upper respiratory tract and
are often grouped into the category of "common cold" viruses. The
most common agents are found with the rhinoviruses and coronaviruses although parainfluenza and adenoviruses also contribute
greatly to the case rate. Pharyngitis, bronchitis, and croup can
be caused by parainfluenza viruses. Table 21.1 provides a summary
of the agents, diseases they cause, and the characteristics of the
disease.

Common colds are transmitted by fomites, direct contact, and
aerosols. Since antibiotics are ineffective, treatment is limited
to alleviating symptoms. Vaccines are not available.

Parainfluenza infections range from inapparent to severe
croup. Most children have been exposed to these viruses and
produced antibodies by age 10. Secretory IgA antibodies seem to be
the most protective against these viruses.

**C.   What bacteria cause lower respiratory infections and what are
the important epidemiologic and clinical aspects of these
diseases?**

Infections of the lower respiratory tract are far less common
but often considerably more severe. Two of the greatest killer
infections in history, tuberculosis and pneumonia, are among these
agents. Table 21.3 provides a complete summary of the agents, the
diseases they cause, and the characteristics of the disease.

Whooping cough or pertussis is a highly contagious, worldwide
disease. The pertussis vaccine (the "P" of DPT) has decreased its
incidence; however, due to the concern about the vaccine's safety,
its use declined in the early 1980's with the result of a significant increase in cases (see Figure 21.5). The vaccine is prepared
from cell wall material of the organism and is difficult to produce
in a completely purified, safe form. However, the number of
serious side effects caused by the vaccine (1/1,000,000 doses) is
greatly outweighed by the numbers of infants left brain-dead due to
the disease.

Classic pneumonia can be lobar or bronchial and is transmitted by respiratory droplets and carriers including those that actually work in hospitals. Most cases of pneumonia occur in individuals with lowered resistance such as the elderly, drug addicts, alcoholics, and those under deep anesthesia. Antibiotics can treat the disease; however, since most patients are already in difficulty, the disease often leaves them in a very weakened state. Chronic ulcerative bronchitis caused by Klebsiella can be very severe; however, lobar pneumonia caused by Streptococcus pneumoniae is more common. Pneumonias caused by Mycoplasmas are generally mild and often referred to as "walking" pneumonia because the patient is ambulatory.

Legionnaires' disease appears to be transmitted by aerosols from contaminated water sources especially improperly installed air conditioning systems. The drug of choice is erythromycin.

Tuberculosis continues to be a worldwide health problem and its incidence in the United States is increasing slowly especially with the newly arrived immigrants, with minorities, and with patients with immune deficiencies such as those with AID'S (see Figure 21.9). It is transmitted by respiratory droplets, and the organisms can persist for years walled off in lung tissue called tubercles. Numerous species can affect humans as illustrated in Table 21.2, some with devastating effects. Isoniazid is the drug of choice. A vaccine is available but is not used in this country since it always causes the recipient to skin test positive for the disease. Health care workers epecially those working with AID'S patients are becoming at risk for this disease.

Psittacosis or ornithosis is transmitted to humans from infected birds. In some states, turkeys and chickens (ornithine birds) are the most common carriers whereas in other states, exotic or imported birds such as parrots and cockatiels (psittacine birds) are more likely sources. The disease is caused by a chlamydian and is usually self-limited but can become serious. Tetracyclines are effective in treatment.

Q-fever is caused by a rickettsian and is transmitted by ticks, aerosol droplets, fomites, and infected milk. It varies from other rickettsial diseases in that it does not cause a rash but does cause a high fever. Treatment is with tetracyclines and a vaccine is available to workers with occupational exposure.

D. **What viruses cause lower respiratory infections and what are the important epidemiologic and clinical aspects of these diseases?**

The most important virus to affect the lower respiratory tract is the influenza virus. The virus has the ability to affect millions by changing its antigenic structure through variations called antigenic shifts and drifts. The virus has surface antigens called hemagglutinins which are responsible for their infectivity and occasionally, neuraminidases, all of which can be changed antigenically to account for their variability **(see Figure 21.14a)**. The disease occurs mainly in December through April and is transmitted in crowded, poorly ventilated conditions especially in older schools and businesses. The disease can be diagnosed by clinical symptoms and by immunological tests. Vaccines can help prevent the disease but must be correlated with the current antigenic variations that are "expected" during that time period.

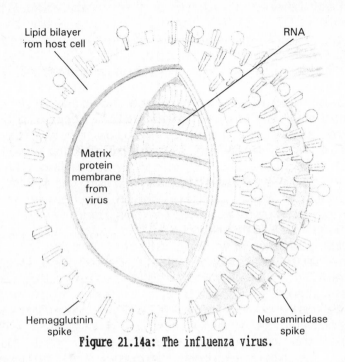

Figure 21.14a: The influenza virus.

The respiratory syncytial and adenoviruses can cause acute respiratory infections which can lead to viral pneumonia. Treatment is for symptoms only.

Table 21.3 provides a summary of the agents, diseases, and the disease characteristics of these viruses.

E.   What fungi cause lower respiratory infections and what are the important epidemiologic and clinical aspects of these diseases?

Most fungal infections occur in immunodeficient and debilitated patients. They are usually transmitted by infectious airborne spores, and some can be treated with amphotericin B. Although all of these diseases affect the respiratory system, most exhibit cutaneous and systemic forms as well. Table 21.3 provides a complete summary of the most important agents, the disease they cause, and the disease characteristics.

Coccidioidomycosis occurs in the warm, arid regions of the desert Southwest, and histoplasmosis is associated with the Mississippi river valley regions. The exposure rates of these diseases can be very high but the morbidity is relatively low. Cryptococcosis occurs wherever there are large infected flocks of birds such as pigeons.

216

Pneumocystis pneumonia occurs primarily in immunodeficient patients, especially those with AIDS. It is diagnosed by finding the organism in biopsied lung tissue.

Lung fluke infections occur primarily in Asia and the South Pacific where infected shellfish are eaten. Thorough cooking can prevent the disease and drugs are available for treatment.

LEARNING ACTIVITIES

Complete each of the following items by supplying the appropriate word or phrase.

1. Another term for sore throat is ___Pharingitis___.

2. Infections of the sinuses are known as ___sinusitis___.

3. Club-shaped Gram-positive rods that form metachromatic granules are ___Diptheroids___.

4. Middle ear infections are called ___Otitis media___.

5. The most common cause of colds seems to be ___Rhinovirus___ followed closely by viruses in the group _____.

6. A high-pitched, noisy respiration is known as ___stridore___ while an acute obstruction of the larynx can result in a barking sound called ___Croup___.

7. Pertussis, a highly contagious disease is also known as ___whooping___ ___cough___.

8. The stage of whooping cough that exhibits the worst symptoms is the ___Paroxysmal___.

9. Failure to keep the airway open during a case of whooping cough may result in a lack of oxygen resulting in ___cyanosis___.

10. Solidified fibrin deposits that form in lobar pneumonia may cause _____.

11. Mycoplasma pneumonia is also known as ___Primary___ ___atypical___ ___pneumonia___.

12. Since patients with Mycoplasma pneumonia often remain ambulatory it is called ___walking___ pneumonia.

13. The disease that affected many of the war veterans attending a convention in Philadelphia in 1976 was _____ disease.

14. A mild form of legionellosis which affected a large number of people in a health department in Michigan was _____ fever.

15. "Cheesy" appearing lesions in cases of tuberculosis are called _____ lesions.

16. Cases of tuberculosis that invade other body tissue forming tiny, millet seed lesions are called _____ tuberculosis.

17. A waxy substance from the cell wall of the mycobacterium that is used in skin testing is called _Tuberculin_.

18. A vaccine for tuberculosis that is made from an attenuated strain is the _BCG_ strain.

19. Psittacosis is also known as _Parrot_ fever.

20. The Q in Q-fever stands for _Qualium_.

21. A surface antigen found on the influenza virus is a _Hemagglutinen_.

22. A type of antigenic variation in flu viruses which is caused by a reassortment of genes is known as _Antigen_ _Shift_.

23. RSV infections cause cells in cultures to form multinucleated masses called _____.

24. <u>Coccidioides</u> <u>immitis</u> forms highly infectious spores called _Arthospores_.

25. Another name for histoplasmosis is _Darling_ disease.

26. The most common cause of pneumonia in AIDS patients is _PCP_.

27. Symptoms that follow the recovery of a disease such as diphtheria are called _____.

28. Another term for the common cold is _____.

29. The word, pertussis, actually means _Violent_ _Cough_.

30. An inflammation of the pleural membranes that causes painful breathing is _Pleurisy_.

31. Tuberculosis was once known as _____.

Match the following terms with the description listed below.

Key Choices:

a. Bacterium   d. Fungus
b. Chlamydian   e. Virus
c. Helminth   f. Rickettsian

___ 1. Causes lobar pneumonia.

___ 2. Causes "Parrot" fever.

___ 3. Are common secondary invaders following a cold but will respond to antibiotics such as penicillin.

___ 4. Causes histoplasmosis.

___ 5. <u>Pneumocystis</u> <u>carinii</u>

___ 6. Responsible for the swine flu epidemic of 1919.

___ 7. Infections may require treatment with amphotericin B.

___ 8. Causes Q-fever.

___ 9. <u>Paragonimus</u> <u>westermani</u>

Match the following terms with the description listed below. Choices may be used more than once.

Key Choices:

a. <u>Streptococcus</u> <u>pyogenes</u>   f. <u>Corynebacterium</u>
b. <u>Streptococcus</u> <u>pneumoniae</u>   <u>diphtheriae</u>
c. <u>Hemophilus</u> <u>influenza</u>   g. Rhinoviruses
d. <u>Mycoplasma</u> <u>pneumoniae</u>   h. Coronaviruses
e. <u>Staphylococcus</u> <u>aureus</u>   i. Parainfluenza viruses

___ 8. Bacterial cause of sinusitis.

___ 9. These bacteria reside in the pharynx but their toxin circulates.

___ 10. Prevented by means of a vaccine.

___ 11. Viral agent that causes the "croup."

___ 12. Common agent to cause "strep throat."

___ 13. Bacterial cause of middle ear infections.

219

_g_ 14. Most common agents to cause "colds."

_f_ 15. Cannot cause disease unless lysogenized.

Match the following terms with the description listed below. Choices may be used more than once.

Key Choices:

a. Coccidioides immitis
b. Histoplasma capsulatum
c. Cryptococcus neoformans
d. Pneumocystis carinii
e. Paragonimus westermani
f. Streptococcus pneumoniae

g. Mycoplasma pneumoniae
h. Staphylococcus aureus
i. Orthomyxoviruses
j. Respiratory syncytial virus

k. Adenoviruses
l. Nocardia asteroides
m. Bordetella pertussis
n. Klebsiella pneumoniae
o. Legionella pneumophila
p. Mycobacterium tuberculosis

q. Mycobacterium bovis
r. Chlamydia psittaci
s. Coxiella burnetti

_i_ 16. Undergoes antigenic shift and drift.

_m,p_ 17. Can be prevented by means of a vaccine.

_r_ 18. Contracted from nonquarantined, infected tropical birds.

_k_ 19. Disease caused by an organism that lacks a cell wall.

____ 20. Extremely severe bacterial infection affecting chronically ill, elderly patients.

____ 21. Forms granulomas or tubercles in the lungs.

____ 22. Contracted from guano-infected cave dust.

_m_ 23. Causes whooping cough.

_o_ 24. Soil bacterium that could be aerosolized and transmitted through air conditioners and humidifiers.

_p_ 25. Transmitted through unpasteurized milk.

_c_ 26. Capsulated yeast transmitted in pigeon droppings.

_d_ 27. Primarily affects patients with immune deficiencies.

_a_ 28. Causes Valley Fever, common to desert Southwest.

REVIEW QUESTIONS

True/False (Mark T for True, F for False)

__T__ 1. Prompt treatment of cases of strep throat is important to avoid the possibility of rheumatic fever.

__F__ 2. Swimmers often contract infections of the ear caused by *Pseudomonas* species of <u>Hemophilus</u> due to its resistance to chlorine.

__F__ 3. Since the case rate of whooping cough has declined, there is little need to require the vaccine.

__T__ 4. Chronic alcoholics, drug addicts, and the elderly are at great risk to a bronchial pneumonia caused by <u>Klebsiella</u>. *causes GI*

__F__ 5. The vaccine for tuberculosis is not used in this country because there are only a few hundred cases per year. *→ recipient develop + skin test make difficult to know who has it*

__T__ 6. <u>Pneumocystis</u> pneumonia is one of the most common fungal agents to affect immunodeficient patients, especially those with AIDS. *Reduce immune defense.*

Multiple Choice

__C__ 7. In cases of tuberculosis:
   a. only the chronically ill are at risk.
   b. the causative agent produces a potent exotoxin.
   c. the bacteria are phagocytized but not killed.
   d. the disease develops rapidly and is usually fatal.

__b__ 8. The orthomyxoviruses are characterized by all the following except:
   a. the viruses possess hemagglutinins and neuraminidases which are necessary for infectivity.
   b. with each antigenic drift, the virus changes completely and would result in a major epidemic.
   c. most viruses in this family are classed as A,B,or C.
   d. amantadine seems to block influenza A replication.

__b__ 9. Respiratory infections caused by fungi:
   a. are extremely rare in this country.
   b. can be detected with skin tests.
   c. occur only in the immunocompromised.
   d. are best treated with antibiotics such as penicillin.

__d__ 10. Q-fever:
   a. is caused by a chlamydian.
   b. has a very high mortality rate.
   c. causes a high fever and severe rash.
   d. can be transmitted through direct contact or by consumption of raw milk.

CLINICAL PERSPECTIVES

The following clinical case was based on an actual case presented in the Morbidity Mortality Weekly Report (MMWR) published by the Centers for Disease Control (CDC). Some of the details have been changed to better illustrate the type of data that may be obtained from the patient. This case should illustrate how epidemiological, clinical, cultural, and other types of information can be used to develop a complete diagnosis and treatment of a disease.

Following the presentation of the case, a series of questions will be made relative to the diagnosis, suggested laboratory tests and type of therapy required. Try to answer these yourself before reading the actual answers. At the end of the case, a short follow-up analysis and additional comments will be made to help correlate all the information concerning this disease.

CASE HISTORY

A 30-year-old woman was admitted to a community hospital in Fargo, North Dakota, on September 10 with a one-week history of fever of 102°F, headache, myalgia, three days of nonproductive cough, and shortness of breath. Radiologic evidence suggested pneumonitis with bilateral lower lobe consolidation. Patient indicated that she attended a state fair four days prior to the onset of symptoms and that she had visited several displays of animals which included pigs.

Preliminary diagnosis?

Pneumonitis, possible influenza.

Suggested laboratory tests?

Bacterial, fungal, and viral cultures. Serological and monoclonal antibody tests for influenza.

Laboratory report:

Bacterial and fungal cultures negative, virus culture positive. Monoclonal antibody test for influenza A positive.

Probably diagnosis?

Influenza, Type A. Serotype not determined.

Suggested treatment?

Broad-spectrum antibiotics for possible secondary infection, ventilation and other respiratory support.

Follow-up and analysis:

Despite intensive support, her respiratory failure persisted, and the patient died on September 18. Postmortem samples were taken and sent to the CDC. Their results confirmed the diagnosis of influenza A and also found it positive to influenza A (H1N1) which is related to swine influenza. Apparently the victim had contracted the virus from an infected hog at the state fair. Veterinarians working at the fair noted than an influenza-like illness had occurred among the pigs at the fair. No other human cases were reported.

Influenza A viruses circulate naturally among many nonhuman hosts including swine, horses, and birds. The virus can act as the possible reservoir from which other strains capable of infecting humans may evolve. Because of their potential variations, some strains can be very severe and, as in this case, lethal.

ANSWER KEY

Fill-in-the-Blank

1. Pharyngitis 2. Sinusitis 3. Diphtheroids 4. Otitis media
5. Rhinoviruses, coronaviruses 6. Stridor, croup
7. Whooping cough 8. Paroxysmal 9. Cyanosis 10. Consolidation
11. Primary atypical pneumonia 12. Walking 13. Legionnaires'
14. Pontiac 15. Caseous 16. Miliary 17. Tuberculin 18. BCG
19. Parrot 20. Query 21. Hemagglutinin 22. Antigenic shift
23. Syncytia 24. Arthrospores 25. Darlings 26. PCP 27. Sequelae
28. Coryza 29. Violent cough 30. Pleurisy 31. Consumption

Matching

1.a 2.b 3.a 4.d 5.d 6.e 7.d 8.f 9.c

8.a,b,c,e 9.f 10.f 11.i 12.a 13.a,b,c 14.g,h 15.f

16.i 17.i,m,p 18.r 19.g 20.f,g,h,n 21.p,q 22.b 23.m 24.o 25.q,s 26.c 27.d,l 28.a

Review Answers

1.   **True**   If treatment is delayed, the organisms could interact with the immune system and give rise to rheumatic fever. (pp. 576-577; Figure 21.1)

2.   **False**   Infections of the ear are commonly caused by species of Pseudomonas that are resistant. (p. 579)

3.   **False** Actually the case rate of whooping cough is increasing in many parts of the country. Much of the increase is due to a concern caused by the vaccine's safety. (pp. 581-583; Figure 21.5)

4.   **True** Since <u>Klebsiella</u> is a common inhabitant of the gastrointestinal tract, it can readily be aspirated into the lungs. this can result in an extremely severe pneumonia. (p. 584)

5.   **False** Although the vaccine is not used in this country, it is because of the problem of the recipient developing a positive skin test which would make it more difficult to determine possible cases. (p. 589)

6.   **True** <u>Pneumocystis</u>, once thought to be a protozoan, causes a severe pneumonia in patients with reduced immune defense systems, especially with AIDS patients. (p. 596)

7.   **c** Tuberculosis can affect anyone; the disease is caused by massive tissue destruction probably due to the reaction to the cell wall material; toxins have not been detected; the disease develops slowly but with treatment can be arrested and the patient can return to a normal life. (pp. 585-589; Table 21.2)

8.   **b** Antigenic drift results from mutations in genes that code for hemagglutinin and neuraminidase. The process occurs slowly and continuously, resulting in a gradual change of antigenicity.
(pp. 591-593; Figure 21.14)

9.   **b** Many respiratory infections such as histoplasmosis and coccidioidomycosis are fairly common in the Southwest; they can affect anyone, regardless of condition, but are most severe in the immunosuppressed; and they are treated with amphotericin B.
(pp. 594-596)

10.   **d** Q-fever is caused by the rickettsian, <u>Coxiella</u> <u>burnetti</u>, is rarely fatal, and causes a fever without a rash.
(p. 590; Figure 21.11)

CHAPTER 22

ORAL AND GASTROINTESTINAL DISEASES

Maintaining good nutrition and appropriate sanitation and hygiene practices are the most effective means of avoiding virtually all of the diseases associated with the oral cavity and gastrointestinal tract. In the United States, most of the populace is indeed fortunate to be able to avoid most of these diseases; however, much of the credit for clean water and proper disposal of sewage must go to the diligent efforts of state and local health departments. Without their efforts, the picture for most of us would not be so rosy.

The problems of obtaining adequate nutrition, clean water, and effective sewage treatment for many other people in this world are painfully made aware to us by observing the underdeveloped countries in Asia, Africa, and elsewhere. Diseases described in this chapter are often the rule rather than the exception. The problems have been so great that one of the primary goals of the World Health Organization for this decade is to ensure clean, disease-free, drinking water to all the people of the world. It is a task that sounds simple to acquire but is enormously difficult to accomplish. This gift of clean water, even though we take it for granted in this country, is one that is still unattainable for millions of people throughout the world. The essay on cholera at the end of this chapter reemphasizes the vital importance of safe, disease free water.

STUDY OUTLINE

I. **Diseases of the Oral Cavity**
    A. Bacterial diseases of the oral cavity
        1. Plaque
        2. Dental caries
        3. Periodontal disease
    B. Viral diseases of the oral cavity
II. **Gastrointestinal Diseases Caused by Bacteria**
    A. Bacterial food poisoning
    B. Bacterial enteritis and enteric fevers
        1. Salmonellosis
        2. Typhoid fever
        3. Shigellosis
        4. Asiatic cholera
        5. Vibriosis
        6. Other bacterial enteritis
    C. Bacterial infections of the stomach, esophagus, and duodenum
    D. Bacterial infections of the gallbladder and biliary tract
III. **Gastrointestinal Diseases Caused by Other Pathogens**
    A. Viral gastrointestinal diseases
        1. Viral enteritis
        2. Hepatitis
    B. Protozoan gastrointestinal diseases
        1. Giardiasis
        2. Amoebic dysentery and chronic amebiasis
        3. Balantidiasis
    C. Effects of fungal toxins
    D. Helminth gastrointestinal diseases
        1. Liver fluke infections
        2. Tapeworm infections
        3. Trichinosis
        4. Hookworm infections
        5. Ascariasis
        6. Other helminth diseases

REVIEW NOTES

A. **What kinds of pathogens cause diseases of the oral cavity, and what are the important epidemiologic and clinical aspects of these diseases?**

Diseases of the oral cavity can be caused by bacteria and viruses and include dental caries, the most common infectious disease in developed countries. The bacteria that cause disease are commonly found in the oral cavity. Most diseases can be prevented by proper oral hygiene and the use of fluorides and dental sealants. Table 22.1 provides a summary of the oral diseases, the causative agents, and the disease characteristics.

Dental plaque is a continuously formed coating of micro-organisms and organic matter on tooth enamel. Plaque formation is the first step in the process of tooth decay and gum disease. Three factors seem to be important for caries to develop: (1) the presence of caries bacteria, (2) a diet containing relatively large amounts of sugar, and (3) poor oral hygiene or poorly calcified enamel. <u>Streptococcus mutans</u> is considered to be the primary causative agent of dental caries. It converts sucrose to sticky dextran material and organic acids which eat away the tooth enamel. **See Figure 22.1(b) below** for an illustration of this process.

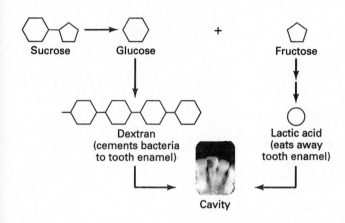

Figure 22.1(b): Dextran formation.

Periodontal disease is caused by a variety of bacteria but especially anaerobic species of <u>Bacteroides</u>. It can be prevented by reducing plaque buildup; by proper flossing; and by treatments of peroxide-sodium bicarbonate, mouth rinses, and, as a last resort, surgery. Acute necrotizing ulcerating gingivitis is the most severe form of gingivitis.

The only important viral disease of the oral cavity is mumps caused by a paramyxovirus. It occurs worldwide, mainly in children, but it can be prevented by a vaccine (part of the MMR series). The disease in adults can be more severe, affecting the reproductive organs.

B. **What bacteria cause gastrointestinal diseases, and what are the important epidemiologic and clinical aspects of these diseases?**

Numerous bacteria affect the gastrointestinal tract. They can be grouped into four categories: those that cause food poisonings; those that cause enteritis and enteric fevers; those that affect the stomach, esophagus, and duodenum; and those that infect the gallbladder and biliary tract. Table 22.2 provides a complete list

of all these agents, the diseases they cause, and the characteristics of each disease.

Bacterial food poisoning is caused by ingesting food
containing preformed toxins. Since the tissue damage is due to the
action of the toxin, it is considered an intoxication or poisoning
rather than an infection. Food poisonings can be prevented by
sanitary handling of the food and proper cooking and refrigeration.

The most common type of food poisoning is staphylococcal
enterotoxicosis. It usually occurs from eating poorly refrigerated
foods, especially dairy and poultry products. The agent usually
enters food from human carriers who do not follow proper food-
handling procedures. Several other agents cause food poisonings
usually arising from eating undercooked meats and gravies, dairy
products, and other foods. Clostridial food poisoning is quite
common and causes abdominal cramps, diarrhea, and nausea following
the consumption of red meats, turkey, and gravies that have not
been properly refrigerated.

Enteritis is an inflammation of the intestines while enteric
fevers are systemic diseases caused by pathogens that invade other
tissues. All enteritises and enteric fevers are transmitted via
the fecal-oral route and can be prevented by good personal hygiene,
proper chlorination of drinking water, and good sanitation
procedures.

Salmonellosis is often considered a food poisoning because it
can be contracted from improperly prepared foods such as chickens
and turkeys. It is generally self-limited and should not be
treated with antibiotics. Typhoid fever is far more severe and is
caused by Salmonella typhi. It can be diagnosed by the Widal test
and is treated with chloramphenicol. Victims often remain carriers
of the disease for years. A vaccine is available but offers
limited protection. Shigellosis can be contracted from infected
drinking water and can be treated with antibiotics. A more severe
tropical form is known as bacillary dysentery. Vaccines are of
limited help. Asiatic cholera occurs in Asia and Africa in regions
with poor sanitation and inadequate, unchlorinated water.

Treatment requires fluid and electrolyte replacement with
antibiotics. Vaccines are of very limited use. Vibriosis tends to
be more mild and is common where raw seafood such as oysters is
eaten. New strains of vibrios have become lethal. Traveler's
diarrhea can be caused by a wide variety of organisms but most
commonly by enterotoxigenic strains of Escherichia coli. Most
cases are debilitating but self-limited.

Recent studies have indicated a bacterial cause of peptic
ulcrs, chronic gastritis, and other problems of the stomach and
duodenum. Helicobacter pylori (Figure 22.9) is detected with most
patients having these ulcers and following antibiotic therapy, the
symptoms rapidly improve.

Infections of the gallbladder and biliary tract are rare
because bile destroys most organisms with lipid envelopes. The
typhoid bacillus is resistant to bile and can live in the
gallbladder and be released in the feces without causing symptoms.
Victims of the disease may require surgical removal of the

228

gallbladder. Gallstones blocking the ducts can cause infections, usually by _Escherichia coli_ which can then spread to the blood-stream and to other organs.

C. What other kinds of pathogens cause gastrointestinal diseases, and what are the important epidemiologic and clinical aspects of these diseases?

Several groups of viruses, protozoans and helminths cause gastrointestinal diseases. A few groups of fungi can also produce toxins that cause poisonings. Most of these agents are described in Table 22.4 along with the diseases they cause and the charac-teristics of the disease.

Most viral gastrointestinal disorders arise from contaminated water or food and are transmitted via the fecal-oral route. The viruses causing hepatitis B, C, D, and E are transmitted parent-erally from contaminated blood and other body fluids.

Rotavirus infections cause severe diarrhea and kill many children in underdeveloped countries. Replacement of lost fluids is essential. Hepatitis A (infectious) is usually contracted from infected food handlers but can be prevented by means of a gamma globulin shot. A vaccine is being developed. Hepatitis B (serum) is much more severe and requires extended bed rest and a nutritious diet. Medical and dental personnel are at greatest risk to hepatitis B and should take advantage of the vaccine. For a complete list and description of these and all the other viruses that cause hepatitis, see Table 22.3.

Protozoan gastrointestinal diseases arise from cyst contam-inated food and water via the fecal-oral route and can be prevented by drinking uncontaminated water, good personal hygiene, and using proper sanitation. Diagnosis is based on the observation of the cysts in the concentrated stool and most can be treated with antiprotozoan agents.

Giardiasis is common in children but especially with backpackers since the parasite is maintained in raccoons, deer, and other wild animals. It is one of the most identifiable agents because of its paired nuclei or "eyes" as shown in Figure 22.12. Amoebic dysentery and chronic amebiasis occur worldwide and in some regions are so common that they are considered the rule rather than the exception. Balantidiasis also occurs worldwide but mainly in tropical areas. It is a large parasite (see Figure 22.14) and is the only important ciliate to cause disease. Cryptosporoidiosis seems to occur only in immunodeficient patients such as those with AIDS.

Gastrointestinal diseases caused by helminths are mainly acquired in tropical regions and include several kinds of fluke, roundworm, and tapeworm infections. The flukes have interesting life cycles that usually involve snails and crustaceans. They gain access to the host following consumption of vegetation that contains the snails or the larva. Tapeworms are usually contracted by consumption of infected undercooked meats. Roundworms have complex life cycles in which they travel from the intestines

229

following consumption of their eggs, to the heart and lungs and then back to the intestines where they develop into adults. Some, such as the hookworms, can burrow through the skin to cause creeping eruptions. Most worm infections are especially severe in young children. They can be diagnosed by finding the eggs in the stool and can be prevented by good sanitation, by avoiding contaminated water and soil, and by thoroughly cooking foods that might be contaminated.

Fungal toxins can be very poisonous and can result in severe injury and even death. Aflatoxins are potent carcinogens produced by species of <u>Aspergillus</u>. Humans and animals can ingest them from moldy grains and raw peanuts. <u>Claviceps</u> <u>purpura</u> produces a highly toxic granule called ergot which forms in contaminated grains. Several species of wild mushrooms especially <u>Amanita</u> (that are inadvertently picked for dinner) produce a highly toxic poison that causes vomiting, diarrhea, liver damage, hallucinations and even death.

LEARNING ACTIVITIES

Complete each of the following items by supplying the appropriate word or phrase.

1. A coating of microorganisms and organic matter that adheres tightly to the tooth enamel is called _____.

2. Following tooth brushing, a film of glycoproteins from saliva called _____ forms on the surface of the teeth.

3. Another word for tooth decay is _____ _____.

4. A test for caries susceptibility involving saliva and a culture medium could be the _____ test.

5. An ion that is added to drinking water to help toughen the teeth against decay is the _____ ion.

6. A dental sealant that is applied to the teeth to prevent attack by bacteria is _____.

7. A mild form of periodontal disease is _____.

8. ANUG is also known as _____ _____.

9. Inflammation of the testes which can occur with mumps is called _____.

10. A strain of <u>Escherichia</u> <u>coli</u> that produces a severe toxin that affects the gastrointestinal tract.

11. An exotoxin that specifically affects the gastrointestinal tract is an _enterotoxin_.

12. A polynesian food poisoning associated with a coconut delicacy is _____ disease.

13. An inflammation of the intestine is called _Enteritis_.

14. An inflammation of the large intestine with severe diarrhea, mucus, pus, and even blood is _Dysentery_.

15. Invasion of the intestinal tissue and the blood caused by some species of <u>Salmonella</u> is _Enterocolis_.

16. A famous carrier of typhoid fever in the 1900's was _Typhoid_ _Mary_ _Mallon_.

17. A watery, tissue-laden stool produced in severe cases of cholera is _Rice_ _Water_ _Stool_.

18. An especially severe strain of cholera with a slower onset is _____.

19. An intestinal disorder contracted by people who travel extensively is _Traveler's diarrhea_.

20. Hepatitis A is commonly called _Infectious_ hepatitis while hepatitis B is called _Serum_ hepatitis.

21. Hepatitis C is usually transmitted _Parenterally_.

22. An inflammation of the liver caused by viruses or other agents is called _Hepatitis_.

23. A dysentery caused by a large ciliated protozoan is _Balantidium coli_.

24. A fungal toxin produced by species of <u>Aspergillus</u> is _Aflatoxin_.

25. A poisoning contracted by eating grains of rye or wheat contaminated with <u>Claviceps</u> is _____.

26. <u>Fasciola</u> <u>hepatica</u> is commonly known as the _Sheep_ _liver_ _Fluke_.

27. Bladder worms that are found in human tissues are called _____.

28. Cysts of certain tapeworms that form in tissues are called _____ cysts.

29. An allergic reaction to toxins produced by certain flukes is known as _____ _____ .

30. The head end of tapeworms is called the ___ scolex ___ .

31. The roundworms of <u>Ancylostoma</u> and <u>Necator</u> are also called ~~Hookworms~~ .

32. Larvae of hookworms that cause creeping eruptions following their entrance into the body are called _____ _____ .

33. The <u>Trichuris</u> worms are known as _whipworms_ due to their long esophagus.

34. Larvae that reenter the host following deposition and hatching of eggs is called _Enterobius vermicularis_

Match the following terms with the description listed below. Choices may be used more than once.

Key Choices:

a. <u>Staphylococcus aureus</u>
b. <u>Clostridium perfringins</u>
c. <u>Clostridium botulinum</u>
d. <u>Salmonella typhi</u>
e. <u>Salmonella typhimurium</u>
f. <u>Shigella flexneri</u>
g. <u>Vibrio cholerae</u>
h. <u>Vibrio parahaemolyticus</u>
i. <u>Escherichia coli</u>
j. <u>Campylobacter jejuni</u>
k. <u>Yersinia enterocolitica</u>
l. <u>Helicobacter pylori</u>

_g_ 1. Patients with this disease can lose over 20 liters of fluids and electrolytes per day.

_h_ 2. Contracted from eating raw shellfish, especially oysters.

_a_ 3. The most common agent of food poisoning.

_c_ 4. A poisoning caused by improperly canned low-acid foods.

_f_ 5. Causes bacillary dysentery.

_d_ 6. Typhoid Mary transmitted this organism.

_k_ 7. Causes symptoms similar to appendicitis.

_e,f,g,d,k_ 8. Transmitted in infected water.

_j, g, h_ 9. Requires treatment based on antibiotic therapy but most importantly, replacement of body fluids due to copious diarrhea.

232

___L___ 10. May be responsible for peptic ulcers.

___i___ 11. The most common cause of traveler's diarrhea.

___b___ 12. Can be transmitted in very rich meat gravies.

___e___ 13. Baby chicks and pet turtles are not sold to the public because of transmission of this agent.

Match the following terms with the description listed below. Choices may be used more than once.

Key Choices:

    a. Plaque   d. ANUG
    b. Dental caries   e. Mumps
    c. Periodontal disease   f. Pellicle

_____ 14. Primarily caused by <u>Streptococcus</u> <u>mutans</u>.

_____ 15. Inflammation of the parotid gland.

_____ 16. A non-bacterial film of polysaccharides from the saliva.

_____ 17. Inflammation of the gums and loosening of the teeth; probably caused by <u>Bacteroides</u> <u>gingivalis</u>.

_____ 18. An acute severe inflammation of the gingiva with very rapid onset.

_____ 19. A mixture of bacteria and organic material that adheres to the tooth.

Match the following terms with the description listed below. Choices may be used more than once.

Key Choices:

    a. Rotaviruses   j. Amoebic dysentery
    b. Hepatitis A   k. Balantidiasis
    c. Hepatitis B   l. Cryptosporidiosis
    d. Hepatitis C   m. <u>Fasciola</u> <u>hepatica</u>
    e. Giardiasis   n. <u>Clonorchis</u> <u>sinensis</u>
    f. <u>Ascaris</u> <u>lumbricoides</u>   o. <u>Enterobius</u> <u>vermicularis</u>
    g. <u>Trichuris</u> <u>trichiura</u>   p. <u>Taenia</u> <u>solium</u> & <u>saginata</u>
    h. <u>Strongyloides</u> <u>stercoralis</u> q. <u>Trichinella</u> <u>spiralis</u>
    i. <u>Echinococcus</u> <u>granulosis</u>   r. <u>Ancylostoma</u> & <u>Necator</u>

___a___ 20. The most common agent of viral enteritis.

___b___ 21. A virus often transmitted in restaurant food from an infected food handler.

233

_f, g, i, o, f_ 22. Contracted by consuming their eggs.

_i, g_ 23. Causes most damage by forming cysts in tissues.

_o_ 24. Causes pinworms.

_g   c_ 25. Vaccine available to prevent disease.

_____ 26. Larvae are coughed up from the lungs to reenter intestines and develop into the adults.

_e_ 27. Contracted by drinking pond water contaminated by animals such as raccoons; common intestinal infection with back-packers.

_____ 28. Helminthic parasites that primarily affect the gastro-intestinal tract.

_p, g_ 29. Contracted from eating meat infected with their cysts.

_j, k_ 30. Causes disease by ulcerating the intestinal mucosa.

_____ 31. Contracted by drinking untreated contaminated water.

_b, c, d_ 32. Can affect the liver.

_____ 33. Larvae penetrate the skin.

_c, d_ 34. Contracted from blood transfusions or unsterile needles.

_m, n_ 35. Are classified as flukes.

REVIEW QUESTIONS

True/False (Mark T for True, F for False)

_____ 1. By age 50, nearly everyone has a form of periodontal disease.

_____ 2. Dental caries is the most common infectious disease in developed countries.

_F_ 3. In cases of bacterial food poisoning, symptoms usually are delayed for a few days until the organisms produce enough toxin. Preformed toxin food - ingested

_F_ 4. Antibiotics should always be given in cases of salmonella food poisonings to reduce the threat of carriers.

_____ 5. Peanut butter and grains such as rye and wheat could be infected by phallotoxin-producing fungi called Amanita.

__T__ 6. Cases of trichinosis have been traced to the consumption of poorly cooked pork, venison, and even horse meat.

__T__ 7. Ascarid worms have been known to wander inside the body to invade other organs and have also crawled up the esophagus and out the mouth.

Multiple Choice

__b__ 8. Select the most correct statement concerning bacterial food poisoning.
   a. Most cases can be treated with antibiotics.
   b. Staphylococcal food poisons cannot be detected by odor or taste and are heat resistant.
   c. Symptoms are usually delayed several days.
   d. <u>Clostridium</u> <u>perfringins</u> produces a food poisoning that is highly lethal.

__a__ 9. In cases of hepatitis:
   a. exposure to hepatitis A may require shots of antibiotics.
   b. both hepatitis B and C can be prevented by means of vaccines.
   c. Although hepatitis A and B are caused by viruses, the hepatitis C variety is actually caused by a species of <u>Hemophilis</u>.
   d. Hepatitis B only occurs in drug addicts.
   e. None of the above are true.

____ 10. Amoebic dysentery is characterized by all the following except:
   a. the disease is transmitted by cysts.
   b. metronidazole is used in treatment.
   c. the parasites cause severe diarrhea but do not invade other tissues.
   d. the asexually reproducing form is a trophozoite.

__d__ 11. Which of the following viruses is not usually implicated in causing cases of enteritis?
   a. rotaviruses   c. echoviruses
   b. Norwalk viruses   d. arenaviruses
                           e. parvoviruses

CLINICAL PERSPECTIVES

The following clinical case was based on an actual case presented in the Morbidity Mortality Weekly Reports (MMWR) published by the Centers for Disease Control. Some of the details have been changed to better illustrate the type of data that may be obtained from the patient. This case should illustrate how epidemiological, clinical, cultural, and other types of information can be used to develop a complete diagnosis and treatment of a disease.

Following the presentation of the case, a series of questions will be asked relative to the diagnosis, suggested laboratory tests and type of therapy required. Try to answer these yourself before reading the actual answers. At the end of the case, a short, follow-up analysis and editorial comments will be provided to help correlate all the information about the disease.

CASE HISTORY

On September 6, a 36-year-old man was treated at a medical clinic in Winkler, Louisiana, for profuse watery diarrhea, vomiting, and dehydration. Just prior to his seeking medical attention he had a very abrubt onset of symptoms and had 15 or more bowel movements. The patient was afebrile, was not taking medication, and had not traveled outside the area during the month before onset. He was admitted that evening to the hospital for observation and testing.

Preliminary diagnosis?

Possible food poisoning, gastroenteritis, vibriosis.

Additional case information:

While in the hospital, the patient "remembered" that he had eaten approximately 10 raw oysters three days earlier at a small seafood bar on the coast and noted that some had an off odor.

Suggested laboratory tests?

Stool culture, blood antibody test for vibrio.

Laboratory Report:

Stool culture yielded toxigenic Vibrio cholerae 01, biotype El Tor, serotype Inaba.

Probable diagnosis?

Asiatic cholera infection probably acquired from the raw oysters.

Suggested treatment?

Replacement of electrolytes and tetracyclines.

Follow-up and analysis:

Patient responded to the antibiotics and improved markedly within three days and was discharged on September 10. Epidemiological studies made at the restaurant did not reveal any additional infected raw oysters and failed to uncover the source of the infected oysters. Since oysters are filter-feeders, they can concentrate the disease agent and allow it to persist for many days after harvesting. Although the source was not positively identified, numerous cases have been reported from eating infected raw oysters. Thorough cooking of all seafood especially shellfish is the best method of prevention.

ANSWER KEY

Fill-in-the-Blank

1. Plaque 2. Pellicle 3. Dental caries 4. Snyders 5. Fluoride
6. Methacrylate 7. Gingivitis 8. Trench mouth 9. Orchitis
10. Enterotoxigenic strain 11. Enterotoxin 12. Bongkrek
13. Enteritis 14. Dysentery 15. Enterocolitis
16. Typhoid Mary Mallon 17. Rice water stool 18. El Tor
19. Traveler's diarrhea 20. Infectious, serum 21. Parenterally
22. Hepatitis 23. Balantidiasis 24. Aflatoxins 25. Ergot
26. Sheep liver fluke 27. Cysticerci 28. Hydatid
29. Verminous intoxication 30. Scolex 31. Hookworms
32. Cutaneous larva migrans 33. Whipworms 34. Retrofection

Matching

1.g 2.h 3.a 4.c 5.f 6.d 7.k 8.d,e,f,g j,k 9.g,h,j 10.l 11.i 12.b
13.e

14.b 15.e 16.f 17.c 18.d 19.a

20.a   21.b   22.f,g,i,o,p   23.i,q   24.o   25.c   26.f,h,r   27.e
28.f,g,h,o,p,r 29.p,q 30.j,k 31.e,j,k 32.b,c,d,j 33.h,r 34.c,d
35.m,n

Review Answers

1.   **True**   Nearly everyone is affected by periodontal disease either with the mild form called gingivitis or the severe form called ANUG. (p. 606)

2.   **True** It is the most common disease because the diet contains relatively large amounts of refined sugar. (p. 604)

3.   **False** Food poisoning is caused by the ingestion of food contaminated with preformed toxins. Therefore symptoms usually appear within a few hours. (p. 609)

4.   **False**   Antibiotics should not be given because they could induce carrier states and could contribute to the development of resistant strains. (p. 610)

5.   **False** These food products could be poisoned by a toxin called aflatoxin produced by a species of <u>Aspergillus</u>. <u>Amanita's</u> are poisonous mushrooms. (pp. 622-623)

6.   **True** Trichinosis is caused by a small nematode, <u>Trichinella spiralis</u>, which is often present in the muscle tissue of hogs, deer, and horses. (pp. 625-626; Figure 22.19)

7.   **True** Ascarid worms are exceptionally large (25-35 cm) and are very active. They have been known to cause abscesses in the liver and other organs during their wandering. (pp. 626-627)

8.   **b**   Most food poisonings should not be treated by antibiotics because it may create carrier states; symptoms occur within a few hours; <u>Clostridium</u> <u>perfringins</u> can cause clostridial food poisoning, which is rarely lethal. Botulism is often fatal. (p. 609)

9.   **a**   Hepatitis A exposure requires preventative gamma globulin shots; only hepatitis B and possibly A is prevented by a vaccine; hepatitis C also appears to be caused by a virus; hepatitis B can occur following blood transfusions or by contact with contaminated blood. (pp. 618-620)

10. **c**   Amoebic dysentery parasites can invade other tissues especially the liver and even the lungs. (p. 621)

11. **d**   Arenaviruses cause Lassa fever and a variety of other hemorrhagic fevers. Rotaviruses, Norwalk viruses, echoviruses, and even parvoviruses all are known to cause enteritis in infants or in small animals (parvoviruses). (pp. 603-604)

CHAPTER 23

CARDIOVASCULAR, LYMPHATIC, AND SYSTEMIC DISEASES

Virtually any disease has an adverse effect on the human body. However, those diseases that have the most profound effect would have to be those that affect the cardiovascular system and those that are capable of causing systemic disorders. In no other way is the body so completely involved and dramatically affected. Diseases such as human plague, anthrax, typhus, and malaria have caused such widespread devastation, misery, and death that they have drastically altered the history of entire civilizations. The populations of whole countries have been decimated to the point that it would be difficult to consider them viable entities. Whole armies have been conquered and destroyed, not by cannons and bullets, but by tiny microscopic bacteria. Even with all these horrors, the thought that some of these agents have already become the arsenal of present day armies, makes the prospects of future wars a hideous nightmare.

Fortunately science has, in most cases, found weapons sufficient to control these diseases and perhaps even to eradicate them. Hopefully, in the very near future, we will begin to see realistic and concerted global efforts being placed on the eradication of many of these disease agents rather than on their use as agents of destruction.

STUDY OUTLINE

I. **Cardiovascular and Lymphatic Diseases**
   A. Bacterial septicemia and related diseases
      1. Septicemia
      2. Puerperal fever
      3. Rheumatic fever
      4. Bacterial endocarditis
   B. Parasitic diseases of blood and lymph
      1. Schistosomiasis
      2. Filariasis
II. **Systemic Diseases**
   A. Bacterial systemic diseases
      1. Anthrax
      2. Plague
      3. Tularemia
      4. Brucellosis
      5. Relapsing fever
      6. Lyme disease
   B. Rickettsial systemic diseases
      1. Typhus
      2. Rocky Mountain Spotted Fever
      3. Other rickettsial infections
   C. Viral systemic diseases
      1. Dengue fever
      2. Yellow fever
      3. Infectious mononucleosis
      4. Other viral infections
   D. Parasitic systemic diseases
      1. Leishmaniasis
      2. Malaria
      3. Toxoplasmosis
      4. Other parasitic diseases

REVIEW NOTES

A. **What pathogens cause bacterial septicemia and related diseases, and what are the important epidemiologic and clinical aspects of these diseases?**

Several different types of normal flora bacteria cause septicemia and related diseases especially members of the Streptococci and Staphylococci. Before the advent of antibiotics, these infections were usually fatal. Today, they are still serious and difficult to treat. Diagnosis of septicemia is made by culturing the blood, catheter tips, urine, or other specimens. Table 23.1 lists the most common agents, the disease they cause, and the disease characteristics.

Rheumatic fever is actually an immunologic reaction to constant, untreated infections caused by Streptococcus pyogenes. Antibiotic treatment will not reverse existing damage, but can

240

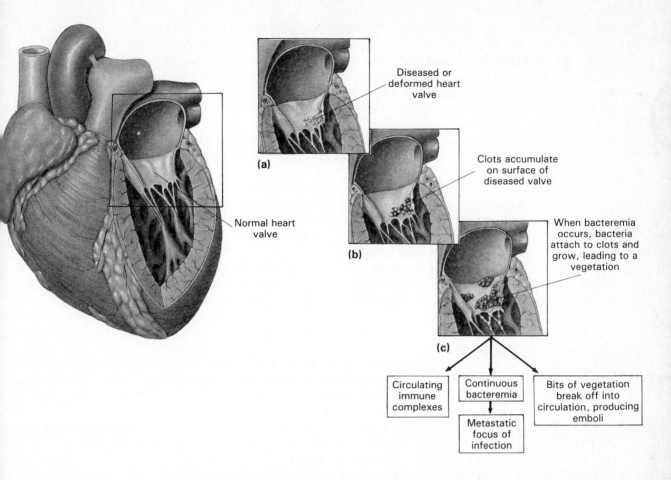

Figure 23.2: Vegetative mass formation on heart valves.

prevent further destruction of the heart valves. Bacterial endocarditis also follows infections caused by species of Streptococci which gain access to the blood following tooth extractions and teeth cleaning. Normally these organisms are removed from circulation immediately; however, if the victim has existing heart valve damage, the bacteria can colonize and eventually form massive vegetations (see Figure 23.2).

B. **What pathogens cause parasitic diseases of the blood and lymph, and what are the important epidemiological and clinical aspects of these diseases?**

Several species of schistosome flukes and a filarial worm cause serious systemic infections. Table 23.1 provides a complete list of the agents, the diseases, and the disease characteristics.

Schistosomiasis or bilharzia is acquired by the aquatic larvae penetrating the skin of a victim and developing in the liver or bladder. The disease is diagnosed by finding the eggs in the stool. Praziquantel can be used for treatment and the disease can be prevented by eradicating the infected snails (rather hard to do) or by avoiding snail-infested water. Most cases occur in Africa,

241

South American, and Asia; however, the disease is prevalent on the island of Puerto Rico. A vaccine has not yet been developed.

Filariasis or elephantiasis is transmitted by mosquitoes and diagnosed by finding the microfilaria in the blood. It is treated with diethylcarbamazine or metronidazole and may require surgical drainage. Avoidance of infected mosquitoes is the best prevention.

C. **What pathogens cause bacterial systemic diseases, and what are the important epidemiologic and clinical aspects of these diseases?**

Several groups of bacteria cause serious systemic diseases, some of which are commonly found in the United States. Table 23.2 provides a complete list of the agents, the disease they cause, and the disease characteristics.

Anthrax is a zoonosis that affects mostly animals and rarely afflicts humans except in countries such as Africa and Asia. It is acquired through spores from infected domestic animals, especially the hides and hair products. U.S. Customs agents have kept infected, imported products from entering this country. The disease is diagnosed by blood cultures or from smears of lesions. Treatment is obtained with antibiotics and the disease can be prevented by vaccinating the animals and reducing the risk of exposure to the agent. Unfortunately, anthrax has been considered as a biological germ warfare weapon.

Plague is also a zoonosis that is maintained in the small animal population throughout the world, including the Western part of the United States as shown in Figure 23.6. It is transmitted by fleas that have fed off of infected rodents and can be diagnosed by stained smears and antibody tests. The disease can be treated with broad spectrum antibiotics but still has a high mortality rate. A vaccine is available for those entering endemic areas.

Tularemia or rabbit fever, is a zoonosis that is maintained in the small animal population and is transmitted by direct contact, inhalation, ingestion, and through insect bites. It can be diagnosed by blood tests and treated with streptomycin. Taxidermists and small game hunters are at great risk for the disease. It can be prevented by avoiding the infected animals and by wearing protective gloves and clothing. The vaccine is not too effective.

Brucellosis or relapsing fever is a zoonosis that is maintained in the domestic large animal population, including cows, pigs, goats, and even dogs. It is an occupational hazard for those in the meat packing industry and is acquired through the skin, by inhalation, by ingestion, and by drinking infected raw milk. It is diagnosed by serological tests, treated with prolonged antibiotic therapy, and prevented by avoiding contact with infected animals and contaminated milk products.

Relapsing fever is a zoonosis maintained in the small animal population and is transmitted by lice and ticks. It can be diagnosed from blood smears, treated with antibiotics, and prevented by avoiding the vectors.

Lyme disease is a zoonosis maintained in the wild animal

population especially deer and is transmitted by ticks. It is diagnosed by clinical signs and by antibody tests. It can be treated with antibiotics such as doxycycline and prevented by avoiding tick bites. The disease appears to be spreading across the country as shown in Figure 23.10, but its increase may be due to better identification of the symptoms.

D. **What pathogens cause rickettsial systemic disease, and what are the important epidemiologic and clinical aspects of these diseases?**

Several species of rickettsia cause serious systemic disease. Rocky Mountain spotted fever and murine typhus are the only two that occur with regularity in the United States. The others occur more often in Europe, Asia, and other parts of the world. The organisms are small Gram-negative, highly infectious, intracellular parasites that can not be cultivated artificially. They are all transmitted by arthropod vectors, including ticks, mites, lice, and fleas. Mosquitoes have not been found to transmit them. They all respond to tetracyclines but even in treated cases, the mortality rate can be high. They are diagnosed by serological tests. Clinical symptoms are all very similar, with a high fever, chills, and a rash being the most common. Table 23.3 provides a complete summary of the diseases, causative agents, area of prevalence, vector, and reservoir.

Murine typhus is transmitted by fleas and is maintained in the rodent population. Southern Texas is the endemic region for the disease.

Epidemic typhus is more of historical note in that it has changed the course of world history several times by infecting armies such as Napoleon's in 1812. It's transmitted by body lice and is common in unsanitary, overcrowded conditions.

Rocky Mountain spotted fever is carried and maintained in the tick and small animal population and is actually distributed more in the Eastern part of the United States than in the Rocky Mountains as noted in Figure 23.13.

E. **What pathogens cause viral systemic diseases, and what are the important epidemiologic and clinical aspects of these diseases?**

Several viruses cause serious systemic diseases, most of which occur with little regularity in the United States. Since antibiotics are of no value, prevention is the best medicine. Vaccines are available for some of these diseases but the best method is to avoid those diseases that are transmitted by vectors. Table 23.4 provides a list of the agents, the diseases they cause, and the disease characteristics.

Dengue or breakbone fever occurs worldwide and is beginning to occur in the United States. It is transmitted by a mosquito and can be diagnosed by serologic tests. Mosquito control seems to be the best means of prevention.

Yellow fever is most common in tropical areas of Central and South America and Africa, is diagnosed by symptoms and serological lab tests, and can be prevented by a vaccine.

Infectious mononucleosis seems to affect those who have not acquired the virus early in life. Children in underdeveloped countries and those that attend public schools seem to have the best opportunity of early exposure and the buildup of antibodies to the virus. Those without this exposure seem to be affected more severely by the virus when they get infected during their teen and young adult years. The disease is diagnosed through clinical symptoms and is only treated symptomatically. Chronic fatigue may be a complication.

Filovirus, bunyavirus, and arenavirus fevers seem to occur predominately in Africa and Central and South America. Parvo-viruses may cause aplastic anemia but most commonly affect dogs and cats. Animal vaccines are available. Coxsackie viruses cause a variety of rare infections of the heart, pharynx, and meninges. There is no treatment, vaccine, or prevention available.

F.   **What pathogens cause parasitic systemic diseases, and what are the important epidemiologic and clinical aspects of these diseases?**

Of all the parasitic systemic diseases, only two, malaria and toxoplasmosis, occur with regularity in the United States. Table 23.4 lists all the diseases, the causative agents, and the disease characteristics.

Leishmaniasis occurs in tropical and subtropical countries where the tiny sandflies are present. A cutaneous and visceral form can occur, with each responding poorly to medication which includes antimony and arsenic. The disease is diagnosed from blood smears or scrapings from the lesions. By reducing the sandfly population and eliminating the rodent reservoirs, the disease could be controlled.

Malaria is the world's greatest public health problem and affects millions worldwide. Most cases in the United States come from people who have been in endemic areas as shown in Figure 23.19; however, with the increase in the number of active carriers, the disease has begun to actually occur in many states, especially in the South. The disease is transmitted by <u>Anopheles</u> mosquitoes and is diagnosed by identifying the parasites in blood smears. Active disease is treated with quinine derivatives; however, some strains are now becoming resistant. Great efforts have been made to find better mosquito-control methods, new forms of chemotherapy, and an effective vaccine. At present, the mosquito and the parasite are winning the battle.

Toxoplasmosis is a zoonosis that is maintained in the rodent population but has infected most domestic cats. The disease agent is now transmitted from the feces of these cats and is most dangerous to pregnant women, especially the fetus, and infants who get in contact with the infected materials. It is diagnosed by finding the parasites in body fluids or tissues and is treated with

244

pyrimethamine and trisulfapyridine. However, once the fetus is infected, the parasites may cause serious damage before it can be treated. Therefore, pregnant women should avoid the "kitty litter" box and any soil contaminated with cat feces.

Babesiosis is not very common but is transmitted by ticks. It can be diagnosed from blood smears and treated with chloroquine.

LEARNING ACTIVITIES

Complete each of the following items by supplying the appropriate word or phrase.

1. A septicemia was also known as _____ _____ due to the spread of organisms through the lymphatics.

2. A drop in blood pressure along with blood vessel collapse due to bacterial infection can result in _Septic_ _Shock_.

3. Puerperal sepsis due to beta-hemolytic streptococci is also known as _streptococcus_ _pyogenes_.

4. Fibrous adhesions that form on heart valves are often called _____.

5. Schistosomiasis is also known as _____.

6. Gross enlargements of limbs caused by blockage of lymphatic vessels may be called _Elephantiasis_. _Wuchereria bancrofti_

7. A respiratory form of anthrax is _Woolsorter's_ disease.

8. Painfully enlarged lymph nodes especially of the groin and armpit are _____.

9. If plague bacteria move from the lymphatics to the circulatory system, they cause a form called _Septicemic_ or _black_ plague because of its hemorrhagic symptoms.

10. The agent that transmits plague from animals to humans is the _Flea_.

11. In ticks, _____ transmission allows the pathogen that is present in the eggs to be transmitted from one generation to another.

12. A septicemia of tularemia can allow the pathogen to cause a _____ _____ which resembles typhoid fever.

13. Infections caused by species of <u>Brucella</u> cause the disease brucellosis, _____ fever, or _____ fever.

14. Since the disease of brucellosis occurs in animals it can be called a __zoonosis__.

15. The epidemic form of relapsing fever is transmitted by a __louse__.

16. Pinpoint-size hemorrhages most common in skin folds that are caused by rickettsia are called _____.

17. Lyme disease is now known to be transmitted by a __tick__.

18. Recrudescent typhus which is a recurrent typhus infection is also known as __Brill Russel__ disease.

19. Scrub typhus is also known as __Tsutsugamushi__ fever and is transmitted by a _____.

20. The vector that transmits Rocky Mountain Spotted Fever is the __tick__.

21. The severe bone and joint pain associated with dengue fever gives reason to call this disease _____ fever.

22. Yellow fever is transmitted by a _____.

23. A cancer caused by the same virus that causes infectious mononucleosis is _____ _____.

24. A parvovirus may be the cause of a symptom of sickle cell anemia called _____ _____.

25. Visceral leishmaniasis is also known as _____ _____.

26. The cutaneous form of leishmaniasis is known as _____ _____.

27. One form of malaria causes large numbers of red blood cells to lyse, resulting in a condition called _____ fever.

Match the following terms with the description listed below. Choices may be used more than once.

Key Choices:

a. <u>Streptococcus</u> <u>pyogenes</u>   d. <u>Schistosoma</u> <u>japonicum</u>
b. <u>Staphylococcus</u> <u>aureus</u>   e. <u>Schistosoma</u> <u>hematobium</u>
c. <u>Streptococcus</u> <u>faecalis</u>   f. <u>Wucheria</u> <u>bancrofti</u>

__b__ 1. Causes toxic shock syndrome and has been linked to super absorbent tampons.

__f__ 2. A parasite that blocks lymphatic channels.

246

_b_e_ 3. Affects the lining of the heart, causing subacute bacterial endocarditis.

_e_ 4. A fluke infection that affects the bladder.

_a_ 5. Causes childbed fever.

_d_ 6. The Aswan Dam allowed the spread of snails which carries the larva of this parasite.

_d,e_ 7. Can be controlled by preventing untreated human waste from entering rivers and streams.

Match the following terms with the description listed below. Choices may be used more than once.

Key Choices:

    a. <u>Bacillus</u> <u>anthracis</u>   d. <u>Brucella</u> species
    b. <u>Yersina</u> <u>pestis</u>   e. <u>Borrelia</u> <u>recurrentis</u>
    c. <u>Francisella</u> <u>tularensis</u> f. <u>Borrelia</u> <u>burgdorferi</u>

_a,d_ 8. A vaccine is available to control the disease in animals.

_a,b_ 9. A vaccine is available for humans.

_b,c,f_ 10. The disease can be prevented by controlling the vectors.

_a_ 11. Endospores of the pathogen can be inhaled, ingested, or can enter a wound to cause infection.

_c_ 12. Has been called rabbit fever and is an occupational hazard for taxidermists.

_a,b_ 13. The pneumonic form is especially fatal.

_d_ 14. Contracted from drinking infected raw milk.

_f_ 15. Causes a bull's-eye rash with symptoms similar to Alzheimers disease.

_b,e_ 16. Causes the formation of buboes.

Match the following terms with the description listed below. Choices may be used more than once.

Key Choices:

    a. <u>Rickettsia</u> <u>prowazekii</u>   e. <u>R.</u> <u>akari</u>
    b. <u>R.</u> <u>typhi</u>   f. <u>Rochalimaea</u> <u>quintana</u>
    c. <u>R.</u> <u>tsutsugamushi</u>   g. <u>Bartonella</u> <u>bacilliformis</u>
    d. <u>R.</u> <u>rickettsii</u>

_a_ 17. The disease that is most responsible for changing the course of history by causing a high mortality in armies.

none 18. Can be prevented by means of a vaccine.

_b_ 19. Transmitted by fleas.

_f, a_ 20. Transmitted by body lice.

_b, d, e_ 21. The reservoir is primarily rodents.

_b_ 22. Most cases appear in Gulf Coast states, especially Texas.

Japanese C 23. Causes scrub typhus because of the vector's association with low scrub vegetation.

_f, g_ 24. Does not occur in the United States.

Match the following terms with the description listed below. Choices may be used more than once.

Key Choices:

a. Dengue fever virus   f. Parvovirus
b. Yellow fever virus   g. Leishmania donovani
c. Epstein Barr virus   h. Plasmodium species
d. Bunyavirus   i. Toxoplasma gondii
e. Arenavirus   j. Babesia microfti

_____ 25. Commonly transmitted by viruses.

_____ 26. Pregnant women need to avoid the cat litter box because of this parasite.

_____ 27. These parasites invade red blood cells in humans.

_____ 28. Transmitted by sand flies and causes Kala azar.

_____ 29. Affects affluent teenagers and often called the "kissing disease."

_____ 30. Vaccines are available to control the disease in dogs and cats.

_____ 31. Causes the hemorrhagic disease called lassa fever.

_____ 32. Known as "breakbone fever."

_____ 33. Best method of prevention is mosquito control.

REVIEW QUESTIONS

True/False (Mark T for True, F for False)

_____ 1. Bacterial endocarditis is generally a life-threatening infection.

_____ 2. Most cases of Rocky Mountain Spotted Fever occur in Montana, Wyoming, and Colorado.

_____ 3. Currently, all malaria parasites respond to the drug, chloroquine, and the disease is declining in most countries, including Africa.

_____ 4. Rheumatic fever is not an infectious disease but actually results from an adverse immune reaction.

_____ 5. Lyme disease is maintained in the wild animal population such as deer and small mammals and is transmitted by a tick.

_____ 6. The Epstein Barr virus appears to exert its effect on lymphocytes.

Multiple Choice

_d__ 7. In the disease of schistosomiasis:
   a. avoidance of mosquitos can reduce the number of carriers.
   b. the parasites are small protozoans that form cysts.
   c. the disease primarily affects small rodents and only recently has affected man.
   d. the parasites form into cercaria to infect their hosts.

_b__ 8. Select the incorrect statement.
   a. Cases of anthrax have occurred from contaminated imported animal hair and skins.
   b. Human plague occurs primarily in Europe and Asia with no cases recorded in the United States for the last 10 years.
   c. Skinning rabbits could result in tularemia.
   d. The agents of brucellosis cause an undulating fever.

_d__ 9. Systemic rickettsial infections are characterized by all the following except:
   a. a skin rash.
   b. respond slowly to antibiotic therapy.
   c. transmitted by arthropod vectors.
   d. have a low mortality rate.

10. Viral systemic diseases:
    a. include dengue fever and yellow fever.
    b. are all lethal.
    c. generally respond to antibiotics.
    d. cannot be prevented by vaccines.

CLINICAL PERSPECTIVES

The following clinical case was based on an actual case presented in the Morbidity Mortality Weekly Reports (MMWR) published by the Centers for Disease Control (CDC). Some of the details have been changed to better illustrate the type of data that may be obtained from the patient. This case should illustrate how epidemiological, clinical, cultural, and other types of information can be used to develop a complete diagnosis and treatment of a disease.

Following the presentation of the case, a series of questions will be asked relative to the diagnosis, suggested laboratory tests, and type of therapy required. Try to answer these yourself before reading the actual answers. At the end of the case, a short, follow-up analysis and editorial comments will be provided to help correlate all the information about the disease.

CASE HISTORY

On July 1, a 45-year-old maintanence worker at a North Carolina textile mill noticed a small, red, pruritic, papular lesion on his left forearm. Over the next several days, the lesion become vesiculated and then developed a depressed black eschar with surrounding edema. On July 7, he sought treatment at a local health facility.

Preliminary diagnosis?

Fungal infection complicated with secondary bacterial infection or cutaneous manifestation caused by a systemic agent such as tularemia, anthrax, or plague.

Additional case information:

Patient was given a topical antifungal agent and oral cephalosporins but on July 8 was hospitalized with worsening edema, pain, fever, and chills. Patient had not traveled outside North Carolina or been exposed to domestic or wild animals and had not worked with objects made of animal material except those at the textile mill. He did not recall any tick or arthropod bite or exposure.

Suggested laboratory tests?

Cultures of blood and wound specimens and serological antibody analyses for tularemia, plague, and anthrax antigens.

Laboratory report:

Culture tests were negative due to previous application of antibiotics but electrophoretic immunotransblot assay of the anthrax antigens demonstrated a high positive titer of 512 to the protective antigens and lethal factor.

Probable diagnosis?

Cutaneous anthrax possibly acquired at the textile mill.

Suggested treatment?

Intravenous ampicillin with oral cephalosporins.

Follow-up and analysis:

The patient's lesion healed with residual local scarring and he was returned to work in August. Epidemiological studies made at the plant revealed the presence of Bacillus anthracis from several samples of West Asian cashmere wool, a sample of Australian wool, and a few samples of surface debris from the storage area. No other cases were reported but a vaccination program for exposed workers was recommended and stringent controls on all products obtained outside the United States were begun.

ANSWER KEY

Fill-in-the-Blank

1. Blood poisoning 2. Septic shock 3. Childbed fever
4. Vegetations 5. Bilharzia 6. Elephantiasis 7. Woolsorters
8. Buboes 9. Septicemic, Black 10. Flea 11. Transovarian
12. Typhoidal tularemia 13. Undulant, Malta 14. Zoonosis
15. Louse 16. Petechiae 17. Tick 18. Brill-Zinsser
19. Tsutsugamushi 20. Tick 21. Breakbone 22. Mosquito
23. Burkitts lymphoma 24. Aplastic crisis 25. Kala azar
26. Oriental sore 27. Blackwater

Matching

1.b 2.f 3.b,c 4.e 5.a 6.d 7.d,e

8.a,d 9.a,b,c 10.b,c,e,f 11.a 12.c 13.a,b 14.d 15.f 16.b,c

17.a 18.none 19.b 20.a,f 21.b,d,e 22.b 23.c 24.f,g

25.a,b,d,h 26.i 27.h 28.g 29.c 30.f 31.e 32.a 33.a,h

Review Answers

1.   **True** Acute endocarditis is a rapidly progressive disease that can destroy heart valves and cause death in only a few days. Subacute endocarditis, although not as rapidly fatal still may cause serious valve damage. (p. 637)

2.   **False** Most cases of Rocky Mountain Spotted Fever actually occur in the Eastern part of the United States, especially in the Appalachian mountains, probably because of a higher tick population and because of a lack of awareness of the disease itself. (pp. 651-652; Figure 23.13)

3.   **False** Many species of Plasmodia are becoming more resistant to quinine derivatives. Furthermore, the disease is still endemic in most tropical areas, especially in Africa, and may be increasing in other countries, even the United States. (pp. 659-660; Figure 23.19)

4.   **True** Rheumatic fever appears to be a sequelae of infections caused by beta-hemolytic Streptococcus pyogenes. There may be genetic predisposition for the disease. (pp. 636-637)

5.   **True** Lyme disease is caused by the spirochaete, Borrelia bourgdorferi. The agent is transmitted by the deer tick, Ixodes dammini. (pp. 647-649; Figures 23.10 and 23.11)

6.   **True** The EBV seems to infect certain kinds of mature B lymphocytes and replicates in the nucleus of the cell. Its DNA is replicated faster than cellular DNA. (pp. 654-655)

7.   **d** Schistosomiasis is contracted from infective aquatic larvae that burrow through the skin; they cause severe liver damage. The disease has caused misery to humans for centuries. (pp. 637-640; Figure 23.3)

8.   **b** Human plague is endemic in many parts of Europe and Asia but still occurs in many rural areas of the Western part of the United States. (pp. 643-645; Figure 23.6)

9.   **d** Systemic rickettsial infections, especially epidemic typhus, have a very high mortality rate. (pp. 649-650)

10.   **a** Many systemic viral diseases are lethal but some such as mononucleosis are self-limited. Viral diseases cannot be treated with antibiotics, and some such as dengue fever and yellow fever can be prevented with vaccines. (pp. 653-654; Table 23.4)

CHAPTER 24

DISEASES OF THE NERVOUS SYSTEM

Diseases affecting the nervous system are among the most dangerous and debilitating that can affect the human body. At the turn of this century, most of these diseases were untreatable and were almost always fatal or at the very least, horribly grotesque. With diligent research and in many cases, fortunate luck, virtually all of these diseases have been controlled with either vaccines or antibiotics. Important discoveries such as the development of the polio vaccines represent classic scientific research efforts even though some of the early trials ended with near disasters.

Only a few diseases such as African sleeping sickness and Chagas disease await important breakthroughs in terms of complete eradication. But, with additional research especially related to insect vector controls, perhaps even these diseases can someday be controlled. As with all aspects of science, some mysteries still abound. As explained in the essay at the end of this chapter, a number of diseases such as Kuru, Creutzfeld-Jacob disease, and "mad cow" disease still defie complete explanation. Discoveries here could, perhaps, provide clues to other unexplained diseases such as Alzheimers and Lou Gehrigs disease.

REVIEW NOTES

A. **What pathogens cause bacterial diseases of the brain and meninges, and what are the important epidemiologic and clinical aspects of these diseases?**

As illustrated in **Table 24.1,** a wide variety of bacteria can infect the meninges, with certain ones affecting each age level. Most cases of meningitis are acute but some are chronic.

TABLE 24.1 Types of bacterial meningitis

Age	Most Frequent Causative Agents	Comments
Newborn (0–2 months)	*Escherichia* coli, other Enterobacteriaceae, *Streptococcus* species	Average mortality about 50%; incidence 40–50/100,000 live births; maternal transmission
Preschool (2 months–5 years)	*Haemophilus influenzae, Neisseria meningitidis*	Maximum incidence 6–8 months; overall incidence 180/100,000 children
Youth and young adult (5–40 years)	*Neisseria meningitidis, Streptococcus pneumoniae*	Sporadic or epidemic
Mature adult (over 40 years)	*Streptococcus pneumoniae, Staphylococcus* species	Sporadic

These agents can be acquired from carriers or from endogenous organisms. Meningococcal meningitis tends to be the most severe, especially during epidemics, and affects primarily adults. Haemophilus meningitis is the most common and affects primarily children. Vaccines are available to protect children from Haemophilus meningitis and to control Streptococcus pneumoniae in the elderly. Antibiotic treatment varies with the causative agent.

Listeriosis can be transmitted by improperly processed dairy products and can cross the placenta to affect the fetus. It can also affect immunodeficient patients.

Brain abscesses can arise from wounds or as secondary infections caused by anaerobes. Antibiotics should be used as early as possible to control the infection or surgery may be required.

Table 24.2 provides a complete summary of the bacterial agents, the disease they cause, and the disease characteristics.

B. **What pathogens cause viral diseases of the brain and meninges, and what are the important epidemiologic and clinical aspects of these diseases?**

Viral diseases of the brain and meninges are usually very severe and often life-threatening. Since antibiotics are of no value, the patient can only be given supportive therapy and hope. Vaccines are available for rabies and for encephalitis but the latter can only be used for horses. Table 24.2 provides a summary of the agents, the diseases they cause, and the disease characteristics.

Aseptic (nonbacterial) meningitis, although not mentioned in the text, is caused by certain enteroviruses and occasionally by the mumps virus. It is usually a severe but self-limited infection since there is little therapy available.

Rabies has a worldwide distribution and is difficult to control because of the large number of small mammals such as skunks and bats that can serve as reservoirs. The disease can be diagnosed by the finding of Negri bodies in neural cells and by the IFAT. The disease can be treated by thoroughly cleaning the bite wound, injecting hyperimmune rabies serum, and by giving the human diploid cell vaccine. It can be prevented by immunizing all pets and people who may be at risk and by avoiding contact with wild animals, especially the very "friendly" or unafraid ones.

Encephalitis is an inflammation of the brain caused by a variety of togaviruses. It is transmitted by mosquitoes, often from horses, and seems to be maintained in the wild bird populations. It can be diagnosed by culturing blood or spinal fluid in cell cultures or mice. The St. Louis variety seems to affect mostly humans whereas the other varieties affect mostly horses.

Herpes meningoencephalitis often follows a generalized herpes infection and is especially severe.

C. **What pathogens cause bacterial nerve diseases, and what are the important epidemiologic and clinical aspects of these diseases?**

Several bacteria specifically affect the central and/or peripheral nervous system by damaging or destroying the nerve cells directly or by blocking neural impulses. These agents are described in Table 24.2, the disease they cause, and the disease characteristics.

Hansen's disease or leprosy affects millions of people worldwide including the United States. In fact, the disease appears to be increasing in the Southern states, especially Texas (see Figure 24.3), probably because of better detection. Many patients are asymptomatic for years after infection and no diagnostic test including the lepromin skin test can detect every case. Dapsone and rifampin are effective in treating most cases so few individuals ever have need to be placed in a "leper" colony. Vaccines are not available. Recently, the Texas armadillo was found to harbor the exact same leprosy bacillus as affects humans. Perhaps the "love affair" with armadillos may need revision.

Tetanus represents a true infection and toxemia since the organisms (usually spores) must enter a deep wound and germinate to develop the disease. It is treated with antitoxin and antibiotics and prevented by means of the tetanus toxoid (the "T" of DPT shot).

Botulism is normally a food poisoning acquired from ingesting foods containing the preformed neurotoxin. It is treated with a polyvalent antitoxin. Most cases of botulism arise from improperly home canned foods, especially those that are neutral in pH. Cases of infant botulism contracted from honey contaminated with the endospores and wound botulism following contamination of a wound have been recorded.

D. **What pathogens cause viral nerve diseases, and what are the important epidemiologic and clinical aspects of these diseases?**

Only a few viruses specifically affect the nerve tissue. The most important one is the poliovirus. Prior to the development of vaccines in the 1950"s, polio was a common and dreaded disease. Today only a few religious groups opposed to immunization and groups of illegal immigrants who have not received the vaccine are at risk for the disease. Diagnosis is made from cultures and by immunologic methods. Treatment alleviates symptoms, but patients with the paralytic strains of the virus may require use of the "iron lung." Both the injectable Salk and the oral Sabin vaccine are available and each has advantages and disadvantages. Worldwide use of the vaccine could eradicate the disease. See Figure 24.9 for an illustration of the effect of the vaccine. A few cases of postpolio syndrome have occurred and may be related to overuse of previously polio weakened muscles.

E. What pathogens cause parasitic diseases of the nervous system, and what are the important epidemiologic and clinical aspects of these diseases?

The trypanosomes are the most important parasites to affect the nervous system. They also do great damage to the cardio-vascular system as well. Both are transmitted by vectors and are difficult to treat. Table 24.2 provides a summary of these parasites, the disease they cause, and the disease characteristics.

African sleeping sickness occurs in equatorial Africa and is transmitted by the tsetse fly. It is diagnosed by finding the trypanosomes in the blood and can be treated with pentamidine and other drugs but they tend to be very toxic. Without treatment, the disease becomes chronic and the victim lapses into a coma and dies.

Chagas' disease occurs from southern United States to all but the southern-most part of South America and is transmitted by various species of triatomid or "kissing" bugs. It is diagnosed by finding the parasites in the blood and does not respond well to any treatment. It is the leading cause of cardiovascular illness in Central and South America.

LEARNING ACTIVITIES

Complete each of the following items by supplying the appropriate word or phrase.

1. An inflammation of the meninges is called _meningitis_.

2. A complication of meningococcal infections where the cocci invade all parts of the body is _____ _____ syndrome.

3. A barrier in the brain to most antibiotics is called the _____ _____ barrier.

4. Diagnosis of rabies can be made on the observation of _negri_ bodies in neural brain cells.

5. A common symptom of rabies in humans is _Hydrophobia_.

6. The form of encephalitis that primarily affects horses is _____ _____ encephalitis.

7. The most severe form of encephalitis is the _____ _____ encephalitis.

8. The preferred name of leprosy is _Hensen's_ disease.

9. The two forms of leprosy are _Lepromatous_ and _Tuberculoid_.

10. The type of vaccine material given to prevent tetanus is tetanus ___Toxoid___.

11. A form of tetanus that affects the umbilical cord of neonates is tetanus _____.

12. Infant botulism seems to be transmitted by feeding infants ___Honey___.

13. The vaccine currently used to prevent polio is the oral ___Sabin___ vaccine.

14. The fly that transmits African sleeping sickness is the ___Tsetse___.

15. American trypanosomiasis is also called ___Chaga___ disease.

16. A condition which affects survivors polio and causes muscle weakness or paralysis is called ___Post polio Syndrone___.

Match the following terms with the description listed below. Choices may be used more than once.

Key Choices:

a. Neisseria meningitidis   h. Polyomavirus
b. Haemophilis influenzae   i. Mycobacterium leprae
c. Streptococcus pneumoniae j. Clostridium tetani
d. Listeria monocytogenes   k. Clostridium botulinum
e. Rabies viruses   l. Polioviruses
f. Encephalitis viruses   m. Trypanosoma gambiense
g. Herpesviruses   n. Trypanosoma cruzi

e,b,e,d 1. Prevented by effective vaccines.

e,f,i,m,n 2. Disease is maintained in the animal population.

j 3. Affects the spinal cord causing muscle contractions.

b 4. The most common bacterial form of meningitis that affects children under five.

m 5. Transmitted by the tsetse fly.

d 6. Contracted by eating unpasteurized cheese; especially severe to pregnant women and the fetus.

i 7. Agent is grown in armadillos and affects primarily the peripheral nervous system.

258

_e_ 8. Requires the use of the IFAT to diagnose the disease. *Rabies*

_K_ 9. May cause three forms: a poisoning, a wound infection, and an infant infection. *Clostridium wound + can food, honey*

_f_ 10. Transmitted by mosquitoes.

____ 11. Treatment only alleviates symptoms.

_J, K_ 12. Form highly resistant endospores.

_a_ 13. Required the military to reduce stress and overcrowding of trainees to reduce the threat of outbreaks.

_h_ 14. JC and BK viruses are members of this group.

_K_ 15. Contracted from improperly canned foods, especially of the low acid foods.

_e_ 16. Affects the spinal cord causing paralysis and may require the use of the "iron lung."

_g_ 17. Transmitted by the triatomid or "kissing" bug.

Match the following terms with the description listed below. Choices may be used more than once.

Key Choices:

a.	Bacterium	d.	Protozoan
b.	Virus	e.	Fungus
c.	Rickettsian	f.	Helminth

_a_ 18. Causes tetanus.

_d_ 19. Causes Chagas' disease.

_a_ 20. Causes listeriosis.

_d_ 21. Causes Hansen's disease.

_b_ 22. Causes polio.

_b_ 23. Causes rabies.

_b, d_ 24. Causes meningitis.

REVIEW QUESTIONS

True/False (Mark T for True, F for False)

_____ 1. Major contributing factors in the epidemics caused by meningococci has been stress, overcrowding, and a decrease in resistance.

_____ 2. The most common cause of heart attacks in Central and South America is the trypanosome of Chagas' disease.

_____ 3. Fortunately, rabid wild animals are easy to spot because they always exhibit symptoms of aggressive behavior, foaming at the mouth, and irregular gait.

_____ 4. Leprosy or Hansen's disease occurs primarily in Asia, Africa, and South America and is only seen in the United States as a result of imported cases.

_____ 5. Symptoms of botulism include respiratory paralysis, slurred speech, and blurred vision.

_____ 6. Listeriosis is a disease that can cross the placenta of pregnant women and infect the fetus.

_____ 7. Virtually all prion associated diseases exhibit a filamentous protein called amyloid and a significantly high inflammatory response.

_____ 8. Because of the abundance of oxygen, most brain abscesses tend to be caused by aerobes.

Multiple Choice

_____ 9. In cases of polio:
   a. the Salk vaccine is most commonly used in the United States.
   b. transmission is by the fecal-oral route.
   c. all patients contracting polio must live in an "iron lung" for the rest of their lives.
   d. the disease is maintained in the wild animal population.

_____ 10. Which of the following is not true concerning rabies.
   a. Bats can actually carry the virus.
   b. A human diploid fibroblast vaccine is currently available and requires only a few injections.
   c. The virus seems to be confined to the North American continent.
   d. The observation of Negri bodies in the brain of a suspected animal is a positive sign of rabies.

260

_____ 11. Diseases caused by trypanosomes:
  a. are easily treated with antibiotics.
  b. do not occur in the United States.
  c. first gain access to the blood stream via insect bites and then travel to affect other tissue.
  d. are now controlled by effective vaccines.

_____ 12. Tetanus:
  a. can be prevented by means of a live attenuated virus vaccine.
  b. affects the neuromuscular junction and blocks neural transmission.
  c. is a fastidious Gram-negative rod which forms an endotoxin.
  d. is commonly found in the soil, especially if its enriched with manure.
  e. causes serious sequelae such as brain damage following the disease.

_____ 13. Select the most incorrect statement concerning encephalitis.
  a. The most reliable method of identification of the viral agent requires cultivation of the virus in tissue cultures.
  b. The main reservoir of most equine encephalitis viruses seems to be the swamp bird population.
  c. The St. Louis encephalitis variety is transmitted mostly between English sparrows, mosquitos, and humans.
  d. Vaccines are available to protect horses but are not generally used on humans.
  e. All of the above are true.

CLINICAL PERSPECTIVES

The following clinical case was based on an actual case presented in the Morbidity Mortality Weekly Reports (MMWR) published by the Centers for Disease Control (CDC). Some of the details have been changed to better illustrate the type of data that may be obtained from the patient. This case should illustrate how epidemiological, clinical, cultural, and other types of information can be used to develop a complete diagnosis and treatment of a disease.

Following the presentation of the case, a series of questions will be asked relative to the diagnosis, suggested laboratory tests, and type of therapy required. Try to answer these yourself before reading the actual answers. At the end of the case, a short, follow-up analysis and editorial comments will be provided to help correlate all the information about the disease.

CASE HISTORY

On Wednesday morning, April 6, an 18-year-old male who resided in Big Springs, Texas, complained of dizziness, double vision, and nausea. He sought treatment from a local physician who diagnosed a possible inner ear infection. He returned home, vomited once, and developed a shortness of breath, slurred speech, and difficulty in walking. An hour later he collapsed, became cyanotic and stopped breathing. An ambulance took him to the emergency room where he was intubated and stabilized. Extraocular weakness was observed, his pupils were fixed and dilated, and he had generalized muscle weakness. Family members indicated the patient had eaten left-over boiled potatoes, a can of creamed corn, and a TV dinner on Tuesday evening.

Preliminary diagnosis?

Possible food poisoning, botulism; drug overdose.

Suggested laboratory tests?

Cerebrospinal (CS) protein analysis, tensilon test, brain scan, serum and mouse inoculation tests for botulinal toxin. Drug residue tests (if necessary).

Laboratory report:

CS protein normal; tensilon test negative; brain scan showed mild edema consistant with hypoxia; mouse inoculation test for Clostridium botulinum toxin, Type A, was positive.

Probable diagnosis?

Botulism, Type A.

Suggested treatment?

Botulinal antitoxin, ventilatory and supportive therapy.

Follow-up and analysis:

Patient made a full recovery following a three-month stay in the hospital along with four months recuperation at home. The patient's food history suggested that foil-wrapped potatoes may have been the source. They were wrapped, boiled, and left unrefrigerated in water overnight. The next day it was reheated in a microwave and consumed along with the corn and TV dinner. Analysis of the foil wrappings from the potatoes was positive for Botulinum Type A toxin. The foil may have provided the anaerobic conditions necessary for C. botulinum to develop. The other food products were negative. The poisoning would have been prevented by properly refrigerating the potatoes after they were cooked.

ANSWER KEY

Fill-in-the-Blank

1. Meningitis 2. Waterhouse-Friderichsen 3. Blood brain
4. Negri 5. Hydrophobia 6. Venezuelan equine 7. Eastern equine
8. Hansen's 9. Tuberculoid, lepromatous 10. Toxoid
11. Neonatorum 12. Honey 13. Sabin 14. Tsetse 15. Chagas'
16. Postpolio syndrome

Matching

1.b,c,e,l,j 2.e,f,i,m,n 3.j 4.b 5.m 6.d 7.i 8.e 9.k 10.f
11.e,f,g,h,l 12.j,k 13.a 14.h 15.k 16.l 17.n

18.a 19.d 20.a 21.a 22.b 23.b 24.a,b,d

Review Answers

1.   **True** Neisseria meningitidis has caused about 3000 cases of
meningitis per year for the least 10 years in the United States.
The disease was the leading cause of death from infectious disease
among service men due in large part to the stress and overcrowding
during boot camp. (pp. 670-671)

2.   **True** Trypanosoma cruzi is the flagellated protozoan that is
credited for causing the greatest number of heart attacks in the
Southern Hemisphere. It is transmitted by the triatomid bug,
common to those areas. It forms pseudocysts which rupture and
cause inflammation and tissue necrosis. (p. 684)

3.   **False** Rabid animals are often difficult to detect since many
animals release the virus in the saliva 3-6 days before exhibiting
symptoms. In addition, many rabid wild animals are often
"friendly" and quite approachable. (pp. 671-673)

4.   **False** Leprosy does occur in the United States and especially
in many of the Southern states. Increases in the reported cases
may be due to better diagnosis. Fortunately most cases respond to
treatment. (pp. 675-678; Figure 24.3)

5.   **True** Botulism is a neuroparalytic disease with sudden onset
and rapidly progressing paralysis. The toxin prevents the release
of acetylcholine, resulting in the symptoms. (pp. 679-680)

6.   **True** Listeria monocytogenes can be transmitted in unpasteurized milk and cheese products. Once ingested, the agent can cross
the placenta of a pregnant woman and infect the fetus, resulting in
spontaneous abortion, stillbirth, or neonatal death. (p. 671)

7. **False** Although amyloid is found in prion associated patients, most lack any inflammatory response. (pp. 685-687)

8. **False** Most brain abscesses are caused by multiple species and anaerobes are just as likely to cause infection as aerobes. (p. 671)

9. **b** The oral Sabin vaccine which contains live attenuated viruses is currently used. Most actual cases are inapparent or mild and nonparalytic. The virus seems to be maintained solely in the human population. (pp. 680-683)

10. **c** The rabies virus seems to have a worldwide distribution but the type of rabies differs in different regions. (pp. 671-673)

11. **c** Trypanosome diseases do not respond to antibiotics but do require somewhat toxic drugs such as pentamine and suramin. Some do not appear to respond well to any drugs. The American trypanosomiasis does occur in the southern part of the United States and successful vaccines have not yet been produced. (pp. 683-684)

12. **d** Tetanus can be prevented by means of the tetanus toxoid. It affects the spinal cord causing muscle spasms but no after-effects if the patient survives. The organism is a Gram-positive endospore forming rod. (pp. 678-679; Figures 24.6 and 24.7)

13. **a** Serological methods are used both during and after illness. Tissue cultures may be used but are often negative even when the disease is present. (pp. 673-674)

CHAPTER 25

UROGENITAL AND SEXUALLY TRANSMITTED DISEASES

Urogenital and especially the sexually transmitted diseases are the most common and among the most serious diseases in our society today but yet these should be the easiest to avoid and control. Abstinence of sexual contact, avoidance of infected sexual partners, and just plain common sense all seem like very simple methods to control these age-old diseases but unfortunately the incidence rates just keep climbing.

With the discovery of penicillin and the powerful broad-spectrum antibiotics, most of these diseases had been controlled and some had even declined dramatically. However, with the emergence of new, highly resistant strains and with a lackadaisical attitude toward sexual practices, many of these diseases have reemerged along with some new and even more frightening ones. The viral diseases such as genital herpes, genital warts, and especially AIDS have defied all our attempts at their eradication. Vaccines which have proved so effective in the control of other serious diseases seem to be of little value in affecting these diseases. At no other time in the history of disease control has the future looked so clouded with concern and despair. Perhaps, as the essay at the end of this chapter suggests, we will evolve into some type of partnership with them. The question is, will we still have time?

STUDY OUTLINE

I. **Urogenital Diseases Usually Not Transmitted Sexually**
   A. Bacterial urogenital diseases
   B. Parasitic urogenital diseases
II. **Sexually Transmitted Diseases**
   A. Bacterial sexually transmitted diseases
      1. Gonorrhea
      2. Syphilis
      3. Chancroid
      4. Nongonococcal urethritis
      5. Others
   B. Viral sexually transmitted diseases
      1. Herpesvirus
      2. Genital warts
      3. Cytomegalovirus

REVIEW NOTES

A. **What bacteria cause urogenital diseases not usually sexually transmitted, and what are the important epidemiologic and clinical aspects of these diseases?**

Urinary tract infections or UTI's are among the most common of all infections seen in clinical practice. They can ascend from the lower urethra to affect the bladder and kidneys or they can descend from systemic kidney infections to affect the bladder. They are diagnosed from urine cultures, treated with antibiotics, and prevented by good personal hygiene and complete emptying of the bladder. Drinking plenty of fresh water can greatly help to reduce these infections. Various organisms can affect this system and are summarized in Table 25.2 along with the disease they cause and the characteristics of the disease.

Urethritis, or inflammation of the urethra, and cystitis, or inflammation of the bladder, are commonly caused by <u>Escherichia coli</u> and others. Prostate infections may result from a urinary tract infection (UTI) and usually can be treated with antibiotics. Dysuria or painful urination is often indicative of an infection and is very common in the elderly (see Table 25.1). Kidney infections also may result from a UTI and are usually more difficult to treat because of renal failure. Infections of the glomeruli of the kidneys may follow streptococcal or viral infections and may result from immune complexes that deposit in the tissue. **Figure 25.2** illustrates how this mechanism can occur. Treatment of the initial agent can reduce the risk of these infections.

Leptospirosis is considered a zoonosis since it is associated with dogs, rodents, and other wild animals. It is transmitted through the animal's urine and can be treated with antibiotics. Pets can be vaccinated to prevent its transmission to humans. Animal control workers and zookeepers are at risk for this disease.

Vaginal infections or vaginitis is usually caused by opportunistic bacteria and other microbes that multiply when the normal vaginal flora is disturbed. <u>Gardnerella</u> vaginitis is quite common and can be diagnosed by finding "clue cells" as shown in Figure 25.4. It can be treated with metronidazole to eradicate the anaerobes and allow restoration of normal lactobacilli.

Toxic shock syndrome most often arises from the use of high-absorbency tampons although it can follow surgeries that were contaminated with the toxigenic strain of <u>Staphylococcus</u>. Whenever a high fever, rash, and drop in blood pressure occur, treatment should be made promptly.

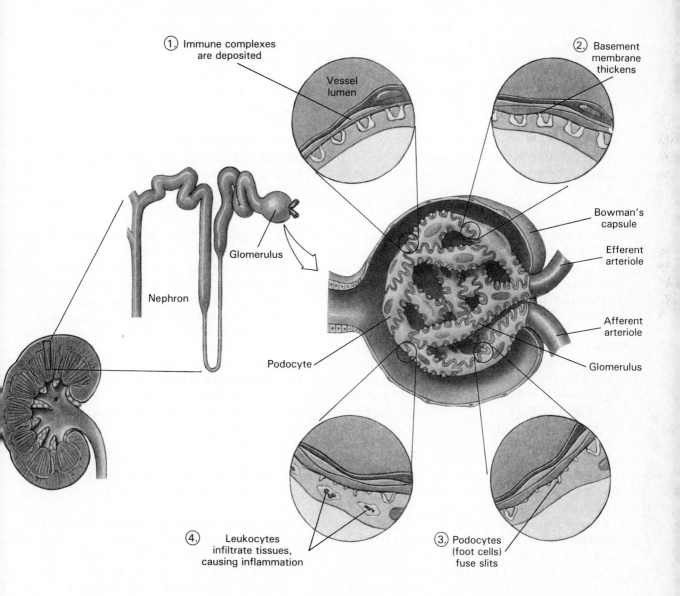

Figure 25.2: Formation of immune complexes in kidneys.

267

B. **What parasites cause urogenital diseases not usually sexually transmitted, and what are the important epidemiologic and clinical aspects of these diseases?**

The most important parasite is <u>Trichomonas</u> which can be transmitted sexually and by indirect contact from linens and contaminated toilet seats. It is a very plump flagellate as shown in Figure 25.5 and is easily diagnosed from smears of the secretions. It can be treated with metronidazole. If the disease is transmitted sexually, both partners must be treated or reinfection will most likely occur. Table 25.2 provides additional information concerning this parasite.

C. **What bacteria cause sexually transmitted urogenital diseases, and what are the important epidemiologic and clinical aspects of these diseases?**

Sexually transmitted diseases or STD's have become an ever increasingly serious health problem in recent years, partly because of changing sexual behavior and because of the development of antibiotic-resistant strains. No vaccines have ever been developed to control any of the STD's and the only methods of prevention are good common sense and avoidance of exposure. Several types of bacteria are responsible for many of the STD's and are summarized in Table 25.2 along with the disease they cause and the disease characteristics.

Gonorrhea is diagnosed by finding Gram-negative intracellular diplococci in pus samples and in culture. It is normally treated with penicillin but the emergence of resistant strains as shown in Figure 25.8 has required the use of other, more potent drugs. Because of the difficulty in recognizing the disease in pregnant women, antibiotic ointments are placed in the eyes of all newborns to prevent possible eye infections. As many as one-half of all infected females will develop pelvic inflammatory disease (PID).

Syphilis has seen a slow, but steady increase as illustrated in Figure 25.9, possibly because of a lack of concern for the disease. The incidence of congenital cases has also increased dramatically. It is diagnosed by immunological tests and is treated with penicillin or other broad-spectrum antibiotics.

Chancroid occurs primarily in underdeveloped countries; however, cases have occurred as a result of prostitutes "visiting" the homes of illegal immigrants. It is diagnosed by finding the Gram-negative bacteria in lesions or buboes and is treated with tetracyclines.

Nongonococcal urethritis or NGU has increased dramatically and appears to be caused primarily by species of <u>Chlamydia</u>. It can be diagnosed by culturing samples from discharges and treated with broad-spectrum antibiotics. Many victims have inapparent infections for years and only later discover that serious damage has occurred to their reproductive organs.

<u>Lymphogranuloma venereum</u> and <u>Granuloma inguinale</u> are venereal diseases most common to tropical and subtropical areas and to

268

Africa and Asia. They are diagnosed by finding the agents in pus or other discharges from lesions or buboes. They are treated with broad-spectrum antibiotics.

D.   **What viruses cause sexually transmitted urogenital diseases, and what are the important epidemiologic and clinical aspects of these diseases?**

Several viruses can be transmitted sexually and because they do not respond to antibiotics, they are often difficult to treat. The most common agents are summarized in Table 25.2 along with the disease they cause and the disease characteristics.

Herpes simplex virus Type 1 (HSV-1) causes most of the oral herpes infections or cold sores and herpes simplex virus type 2 (HSV-2) causes most of the genital infections. Genital herpes is the most common and most severe of the herpes simplex virus infections. It is diagnosed by finding the viruses in vesicular fluids, by immunologic tests, and by clinical symptoms such as those shown in Figure 25.18. Acyclovir can help to reduce the symptoms but can not cure the disease. Once patients are infected the viruses invade the sensory nerves and establish a permanent latency. Reactivation of these viruses may occur spontaneously or by some external influence.

Genital warts has seen an alarming increase in recent years and must be distinguished from cervical dysplasia and carcinoma. Cryosurgery is usually necessary to remove them; however, they often recur.

Cytomegalovirus infections are most severe in fetuses, new-borns, and in the immunocompromised. Monoclonal antibody tests can be used for diagnosis but no effective treatment is available.

LEARNING ACTIVITIES

Complete each of the following items by supplying the appropriate word or phrase.

1.   An inflammation of the urethra is called _____.

2.   An inflammation of the prostrate gland is _____.

3.   UTI stands for ___*Urinary*___ ___*Tract*___ infection.

4.   Women with painful, burning urination, indicative of urethral infections have a condition called ___*Dysuria*___.

5.   An   inflammation   of   the   kidneys   is   known   as _____.

6.   Liver damage and jaundice caused by leptospiras is called ___*Weils*___ disease.

7. Symptomatic epithelial cells covered with bacteria that suggest infections caused by <u>Gardnerella</u> are called _____ cells.

8. Toxigenic strains of staphylococci produce an _____ which enhances the effect of its endotoxin.

9. STD refers to _Sexual_ _Transmiter_ diseases.

10. PMNL's refer to _____ _____.

11. Gonorrhea infections that spread throughout the pelvic cavity indicate a _Pelvic_ _inflammatory_ disease.

12. A hard crusty painless lesion of syphilis is a _Chancre_.

13. Noninfectious granulomatous inflammations seen in tertiary syphilis are known as _Gummes_.

14. The complement fixation or VDRL test for syphilis is named for the _Veneral_ _Disease_ _Research_ _Laboratory_.

15. Notched incisors seen in congenital syphilis are _____ incisors.

16. NGU stands for _Non gonococcal_ urethritis.

17. Self-inoculation with chlamydial infections from the genitals to the eyes could result in _Inclusion_ _Conjonclivitis_.

18. LGV stands for _Lymphoma granule_ _venerium_.

19. Inflamed lymph nodes are called _Bulbes_.

20. The finding of large mononuclear cells called _____ bodies are diagnostic for <u>Granuloma inguinale</u>.

21. Lesions of the mucous membranes of the mouth especially in young children are called _____.

22. Herpes infections of the eye are called herpes _____.

23. A herpetic lesion on a finger can be called a herpes _____.

24. Genital warts is also known as _Condylomas_.

25. A generalized infection in babies caused by the cytomegalovirus is _____ _____ disease.

26. An infection which spreads from the urethra to the bladder would best be termed _____.

Match the following terms with the description listed below. Choices may be used more than once.

Key Choices:

a. NGU   f. UTI
b. STD   g. LGV
c. VDRL   h. HSV-2
d. CMV   i. CID
e. PMNL   j. HSV-1

_h_ 1.   This latent virus causes genital lesions.

_d_ 2.   This virus is especially lethal to neonates.

_c_ 3.   This is a complement fixation test to check for syphilis.

_a_ 4.   This is a chlamydian that causes urethritis.

_b_ 5.   This stands for diseases that are transmitted sexually.

_g_ 6.   This forms pus-filled buboes that must be lanced.

Match the following terms with the description listed below. Choices may be used more than once.

Key Choices:

a. Bacteria   e. Protozoan
b. Virus   f. Helminth
c. Rickettsia   g. Fungus
d. Chlamydia

_a_ 7.   Causes leptospirosis.

_b_ 8.   Causes genital warts.

_a_ 9.   Causes Chancroid.

_a_ 10. Causes Gonorrhea.

_b_ 11. HSV-2.

_e_ 12. Causes trichomoniasis.

271

_a_ 13. Causes toxic shock syndrome.

_d_ 14. Causes nongonococcal urethritis.

_d_ 15. <u>Granuloma inguinale</u>.

_a, g_ 16. Causes pyelonephritis.

Match the following terms with the description listed below. Choices may be used more than once.

Key Choices:

a.	<u>Escherichia coli</u>	i.	<u>Haemophilis ducreyi</u>
b.	<u>Gardnerella vaginalis</u>	j.	<u>Chlamydia trachomatis</u>
c.	<u>Staphylococcus aureus</u>	k.	<u>Calymmatobacterium</u>
d.	<u>Leptospira interrogans</u>	l.	Herpes simplex
e.	<u>Trichomonas vaginalis</u>	m.	Papilloma viruses
f.	<u>Candida albicans</u>	n.	Cytomegaloviruses
g.	<u>Neisseria gonorrhea</u>	o.	<u>Mycoplasma hominus</u>
h.	<u>Treponema pallidum</u>	p.	None of the above

_b_ 17. This causes a frothy, fishy smelling vaginal discharge.

_h_ 18. Forms highly infectious painless chancres and a rash.

_e, f, l, m, n_ 19. Cannot be treated with antibiotics.

_p_ 20. Human infections that can be prevented with a vaccine.

_i, j_ 21. Forms soft, painful, bleeding lesions and enlarged buboes.

_m_ 22. Causes genital warts.

_e_ 23. Causes intense itching with a copious white discharge.

_a_ 24. Most common cause of urinary tract infections.

_c_ 25. Causes toxic shock syndrome.

_a, b, i, k_ 26. Is a Gram-negative rod.

_d_ 27. Is a large spirochaete that is a zoonosis especially in rats and dogs.

_g_ 28. A bacterial STD that has become resistant to many antibiotics.

_f_ 29. Causes a vaginal yeast infection.

_g_ 30. Infected women often are asymptomatic carriers of this agent.

_31. Can cause congenital defects.

REVIEW QUESTIONS

True/False (Mark T for True, F for False)

_F_ 1. Toxic shock syndrome seems to occur only with menstruating women.

_T_ 2. Strains of gonococci which lack pili usually are nonvirulent.

_T_ 3. Once infected, a patient retains herpes simplex viruses for life.

____ 4. CMV infections seem to be most severe in teenagers, causing fever, malaise, and abnormal liver function.

_F_ 5. NGU infections appear to be caused most often by papilloma viruses because of their ease of transmission.

Chlamydia

_T_ 6. Chancroid is caused by a <u>Hemophilus</u> species which forms soft, painful lesions on the genitalia.

_T_ 7. Women are 40 to 50 times more likely to acquire a UTI.

____ 8. Although <u>Trichomonas</u> <u>hominis</u> is considered a commensal, <u>T. tenax</u> causes serious genital infections in cattle and is the leading cause of spontaneous abortion in these animals.

Multiple Choice

_d_ 9. Which of the following venereal diseases is easiest to treat.
   a. Genital warts   c. Tertiary syphilis
   b. HSV-2   d. Gonorrhea

_b_ 10. Syphilis is characterized by all the following except:
   a. the primary stage is characterized by a painless, hard, crusty chancre.
   highly b. the secondary stage is characterized by a noninfectious rash similar to measles.
   c. neurosyphilis can occur in the tertiary stage.
   d. congenital syphilis may cause all types of deformities including saber shin, saddle-nose, and Hutchinson's incisors.

_____ 11. Select the most correct statement concerning UTI's.
   a. UTI's generally are very rare in the United States but rather common in the tropics.
   b. UTI's can rapidly spread throughout the urinary tract by "ascending" or "descending".
   c. Most UTI's cannot be treated with antibiotics.
   d. Once a patient contracts a UTI, they are usually immune to further infections.

_____ 12. In terms of venereal diseases:
   a. Lymphogranuloma venereum is fairly common in the U.S. especially in the Northeast.
   b. Donovanosis is a rare disease and does not respond to antibiotic therapy.
   c. Genital warts seems to have a close link with cervical carcinoma.
   d. CMV infections are only transmitted sexually.

_____ 13. In cases of leptospirosis, all the following are true except:
   a. the disease is a zoonosis.
   b. the spirochaetes are released in the infected animals urine.
   c. the spirochaetes specifically invade the bladder and eventually infect the kidneys of humans resulting in a severe pyelonephritis.
   d. the spirochaetes readily respond to most antibiotics if given early.
   e. a vaccine is available for dogs and cats.

CLINICAL PERSPECTIVES

The following clinical case was based on an actual case presented in the Morbidity Mortality Weekly Report (MMWR) published by the Centers for Disease Control (CDC). Some of the details have been changed to better illustrate the type of data that may be obtained from the patient. This case should illustrate how epidemiological, clinical, cultural, and other types of information can be used to develop a complete diagnosis and treatment of a disease.

Following the presentation of the case, a series of questions will be asked relative to the diagnosis, suggested laboratory tests, and type of therapy required. Try to answer these yourself before reading the actual answers. At the end of the case, a short, follow-up analysis and editorial comments will be provided to help correlate all the information about the disease.

CASE HISTORY

A 36-year-old male presented to the Boston City Hospital sexually transmitted disease (STD) clinic on December 8 with a tender penile ulcer on the glans that had been present for about 2 weeks. This ulcer was accompanied with swollen, tender, left-sided inguinal lymph nodes. He denied having any sexual intercourse while in Massachusetts or during his recent vacation in Florida.

Preliminary diagnosis?

Sexually transmitted disease, possibly syphilis.

Suggested laboratory tests?

Fluorescent treponemal antibody tests, serological tests for syphilis, culture tests, Tzanck smears for herpes simplex.

Suggested treatment?

2.4 million units of procaine penicillin.

Laboratory report:

Negative FTA, negative serology, negative culture tests, negative Tzanck.

Additional case information:

Patient was examined on follow-up study one week later. The ulcer was unchanged and he had developed swollen painful, right-sided inguinal lymph nodes.

Probable diagnosis:

Chancroid

Confirmatory laboratory tests?

Culture test for _Hemophilus ducreyi_, indirect immunofluorescence of ulcer smear with monoclonal antibodies and dot-immunobinding serologic tests.

Recommended treatment?

Oral erythromycin, 500 mg., 4 times a day.

Follow-up and analysis:

Patient improved markedly and serological tests confirmed the presence of _H. ducreyi_ although culture tests were negative. _H. ducreyi_ is often difficult to culture and requires special media and personnel experienced with growing the organism. Isolation and identification rates are often less than 75%. Chancroid is an uncommon, usually tropical disease but outbreaks are known to occur. The organism is highly infectious but usually responds well to treatment.

ANSWER KEY

Fill-in-the-Blank

1. Urethritis 2. Prostatitis 3. Urinary tract 4. Dysuria
5. Pyelonephritis 6. Weils 7. Clue 8. Exotoxin C
9. Sexually transmitted 10. Polymorphonuclear leukocytes
11. Pelvic inflammatory 12. Chancre 13. Gummas
14. Venereal Disease Research Laboratory 15. Hutchinson's
16. Nongonococcal 17. Inclusion conjunctivitis
18. Lymphogranuloma venereum 19. Buboes 20. Donovan
21. Gingivostomatitis 22. Keratoconjunctivitis 23. Whitlow
24. Condylomas 25. Cytomegalic inclusion 26. Urethrocystitis

Matching

1.h 2.d 3.c 4.a 5.b 6.g

7.a 8.b 9.a 10.a 11.b 12.e 13.a 14.d 15.d 16.a,g

17.b 18.h 19.e,f,l,m,n 20.p 21.i,j 22.m 23.e 24.a 25.c 26.a,b,i,k
27.d 28.g 29.f 30.g,j 31.h,l,n,o

Review Answers

1.   **False**   Infections with toxigenic strains of Staphylococcus aureus can affect both males and females and may occur following surgery, boils, or furuncles. (p. 696)

2.   **True**   The gonococci possess sex pili which are essential for conjugation and transmission of antibiotic resistance and common pili which are used for attaching to epithelial cells lining the urinary tract to avoid being swept out with the elimination of urine. (pp. 697-700)

3.   **True**   The HSV viruses appear to become latent in the cells they infect and periodically recur.   This reactivation can be spontaneous or be the result of UV light, stress, or other factors. (pp. 706-708)

4.   **False** CMV infections generally go unnoticed in older children and adults but are most severe and even life-threatening in fetuses and infants. (pp. 711-713)

5.   **False**   NGU appears to be caused by <u>Chlamydia</u> <u>trachomatis</u>. Papilloma viruses cause genital warts. (pp.704-705)

6.   **True**   <u>Hemophilus</u> <u>ducreyi</u> is the causative agent of chancroid or soft chancre to distinguish it from the lesions of syphilis. (p. 704)

7.   **True**   The female urethra is shorter (4 cm) than the male urethra (20 cm) thus greatly increasing the chances of infection. (p. 692)

8.   **False**   <u>Trichomonas</u> <u>tenax</u> and <u>T.</u> <u>hominis</u> are both considered commensals.   <u>T.</u> <u>vaginalis</u> causes serious vaginal infections in humans and <u>T.</u> <u>foetus</u> causes serious vaginal infections in cattle. (pp. 696-697)

9.   **d**   Gonorrhea can be successfully treated with penicillin or spectinomycin. Genital warts and HSV-2 require either surgery or chemotherapy but treatment is not always successful.   Tertiary syphilis is often due to an immune response which cannot be readily treated. (p. 700)

10. **b**   The secondary stage of syphilis is typified by a highly infectious rash which may, in fact, be typical of measles. (p.702)

11. **b**   UTI's are among the most common of infections in the United States and most can be treated with antibiotics.   Since there is a great variety of UTI's, the patient is not immune to subsequent infections. (pp. 692-693)

12. **c**   LGV is rare in the United States with the few cases occurring in Southeastern states. Donovanosis is treatable with ampicillin or tetracyclines. CMV can be transmitted by any body fluid such as saliva, blood, and breast milk, but especially in the urine. (p. 711)

13. **c**   The spirochaetes are transmitted through animal urine and enter the body through mucous membranes.   Once acquired, the spirochaetes travel to the liver where they can cause a serious jaundice. (p. 695)

CHAPTER 26

ENVIRONMENTAL MICROBIOLOGY

Microbes play a most fundamental role in the ecology of all living systems. Without them, in the words of the famous microbiologist, Hans Zinsser, "...the physical world would become a storehouse of well-preserved specimens of its past flora and fauna...useless for the nourishment of the bodies of posterity..." A review of the biogeochemical cycles immediately reveals their essential presence in the conversion of many components of each cycle.

Although the emphasis of this chapter is to provide a description of the role and importance of microbes in our environment, there are still several opportunities to bring in their medical importance as they relate to the air, soil, water, and sewage. Upon completion of this chapter, it should be very clear that as long as life exists on this planet, microbes will continually be an integral part of every aspect of our lives.

STUDY OUTLINE

REVIEW NOTES

A.   What is ecology, and how does energy flow in ecosystems?

Ecology is the study of relationships among organisms and their environment which include interactions of organisms with physical features and with each other. An ecosystem includes all the living or biotic factors and the nonliving or abiotic factors of an environment. Microorganisms can be native or indigenous to the environment or temporary, nonindigenous inhabitants. All the living organisms in an ecosystem make up a community.

Energy in an ecosystem flows from the ultimate source, the sun, to the producers and then to the consumers. Decomposers which are mostly microorganisms obtain their energy from digesting dead bodies and wastes of other organisms. This insures that all nutrients are recycled. See Figure 26.1 for an overview of this process.

B. **Why is recycling important, and how are water and carbon atoms recycled?**

Although energy is continuously available from the sun, nutrients must be recycled to insure their availability. Dead organisms and wastes must be recycled by decomposers as too much matter would be incorporated into this material and life would soon become extinct.

The water or hydrologic cycle insures that water is continuously recycled as all living organisms use this as part of their metabolism. See Figure 26.2 for an overview of this process.

The carbon cycle insures that carbon in both an inorganic and an organic form is always available to living organisms. Carbon is the basic building block for all living organisms. See Figure 26.3 for an overview of this process.

C. **What other biogeochemical cycles exist, and what roles do microorganisms play in them?**

The nitrogen cycle insures that nitrogen in both an inorganic and an organic form is continuously recycled from the atmosphere through various organisms and the soil and back to the atmosphere. All living organisms including microorganisms require nitrogen for their metabolism and thus become essential parts of the nitrogen cycle. **See Figure 26.4**, which illustrates the cycle.

Nitrogen bacteria fall into three categories according to the roles they play in the nitrogen cycle. These are nitrogen fixers, nitrifying bacteria, and denitrifying bacteria.

Nitrogen fixation is the reduction of atmospheric nitrogen to ammonia and is accomplished by free-living aerobes and anaerobes and also by nitrogen fixers in combination with plants called legumes. Rhizobium is an organism that has established a symbiotic relationship with legumes such as soybean and alfalfa to fix atmospheric nitrogen. This helps the bacteria and the plant gain the nitrogen they need. These plants can then be rotated with corn plants that extract nitrogen from the soil.

Nitrification is the conversion of ammonia to nitrites and nitrates. Nitrosomonas converts ammonia to nitrites and Nitrobacter converts the nitrites to nitrates. This helps to provide plants with needed amounts of nitrates which they use in their metabolism.

Denitrification is the conversion of nitrates to nitrous oxide and nitrogen gas. It primarily occurs in waterlogged soils and is a detrimental process for plants because it removes the useful nitrates.

Sulfur is another important element that is continuously recycled as shown in Figure 26.7. Various bacteria carry out important parts of this cycle including sulfate reduction, sulfur reduction, and sulfur oxidation.

Phosphorus also is recycled as illustrated in Figure 26.9. Again, microorganisms become essential parts of the cycle to insure that organic and inorganic components are returned to useful forms.

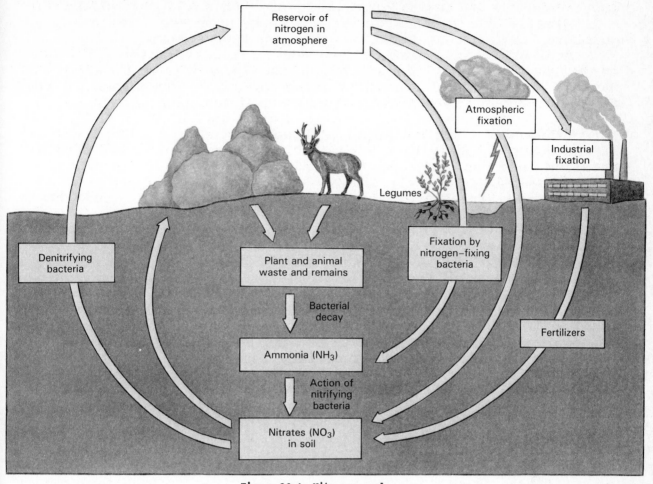

Figure 26.4: Nitrogen cycle.

D. What kinds of microorganisms are found in air, and how are they detected and controlled?

Microorganisms do not grow in air, but spores and vegetative cells can be transmitted by air currents or on dust particles and water droplets. The kinds and numbers of microorganisms vary tremendously in different environments.

Air can be sampled by exposing agar plates to the air or by drawing air over the surface of an agar plate or liquid medium. The organisms can then be identified and studied.

Microorganisms in the air can be controlled by chemical agents such as triethylene glycol, by UV radiation, by filtration, and by unidirectional airflow such as a laminar airflow system.

E. What kinds of microorganisms are found in soil, and what are their roles in biogeochemical cycles and as pathogens?

Soil consists of inorganic and organic components. The inorganic components include rocks, minerals, water, and gases. The organic components include humus and microbes of all taxonomic groups such as bacteria, fungi, algae, viruses, and protists.

Various physical factors affect these microbes. These include moisture, oxygen concentration, pH, and temperature.

Microorganisms also alter the characteristics of their environment as they are effective decomposers and they release wastes, some of which can be used by other organisms. They represent important members of the carbon and nitrogen cycles.

Soil pathogens affect mainly plants and insects and rarely affect humans. The main human pathogens found in soil are members of the genus _Clostridium_. These include those that cause tetanus, botulism, and gas gangrene.

F. How do freshwater and marine environments, and their microorganisms, differ?

Freshwater environments are characterized by low salinity and a great variability in temperature, pH, and oxygen concentration. Because of these extremes, a wide diversity of microorganisms may be found. Most prefer moderate temperatures and a neutral pH environment. Aerobic bacteria are abundant where oxygen is plentiful and anaerobic bacteria accumulate where oxygen is depleted.

Marine environments are characterized by high salinity, and smaller variabilities in temperature, pH, and oxygen concentrations. As the depth increases, pressure increases and sunlight penetration decreases. Microbes from all taxonomic groups are found; however, they must adjust to the higher salinity. Photosynthetic microbes are found near the surface, heterotrophs in the middle and lower strata, and decomposers in the bottom sediments.

G. How do water pollution and waterborne pathogens affect humans?

Water is polluted if a substance or condition is present that renders the water useless for a particular purpose. Thus, the concept of water pollution is a relative one, depending both on the nature of the pollutants and the intended uses of the water. Water may be too polluted for drinking but may be safe for swimming or boating. Table 26.1 lists the effects of water pollution.

Many human pathogens can be transmitted in water. These include _Salmonella_, _Shigella_, _Vibrio's_, _Escherichia_, _Giardia_, Hepatitis A viruses, and many others. See Table 26.2 for a complete list of the human pathogens transmitted in water.

H. How is water purified, and how is it tested to determine purity?

Purification procedures for human drinking water are determined by the degree of purity of the water at its source. If the source is very contaminated, the procedures may be more elaborate. The usual procedures involve flocculation of suspended matter, filtration through beds of sand to remove bacteria, and chlorination to insure the water is safe to drink.

Tests for purity are designed to detect coliform bacteria,

especially <u>Escherichia</u> <u>coli</u> which is always associated with the human intestinal tract. These tests include the multiple-tube fermentation test which is illustrated in Figure 26.15, and the membrane filter method which is shown in Figure 26.16.

I.   What is sewage, and what processes are involved in primary, secondary, and tertiary sewage treatment?

Sewage is used water and the wastes it contains. It usually is about 99.9% water and about 0.1% solid or dissolved wastes which includes household wastes such as human feces, detergents, and grease; industrial wastes such as acids and other chemicals; and wastes carried by rainwater that enters the sewers.

Sewage treatment involves three main processes (**see Figure 26.17**).   Primary treatment is the removal of solid wastes by physical means; secondary treatment is the removal of organic matter by the action of aerobic bacteria; and tertiary treatment is the removal of all organic matter, nitrates, phosphates, and any surviving microorganisms by physical and chemical means.   Most treatment plants utilize only the primary and secondary processes.

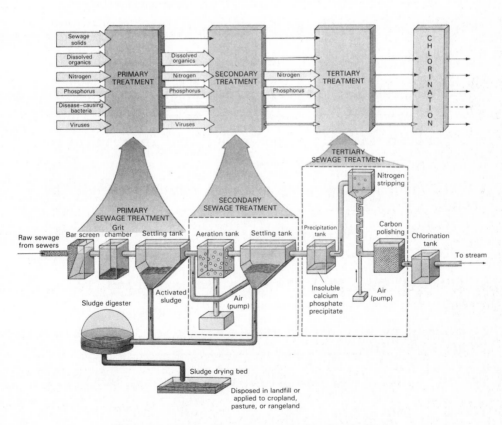

Figure 26.17: Overview of a sewage treatment plant.

LEARNING ACTIVITIES

Complete each of the following items by supplying the appropriate word or phrase.

1. The study of the relationships among organisms and their environment is called _____.

2. Native organisms always found in a given environment are _____.

3. Organisms that can produce energy from the sun are called _____.

4. Microorganisms act as _____ by obtaining their energy from wastes of producers and dead bodies.

5. Mechanisms by which recycling of chemicals such as carbon and nitrogen occurs are termed _____ cycles.

6. The biogeochemical cycle that recycles water is the water or _____ cycle.

7. Atmospheric carbon dioxide and water vapor form a blanket over the earth's surface creating the _____ effect.

8. In the carbon cycle, the atmospheric form is _____.

9. In the nitrogen cycle, the atmospheric form is _____.

10. The process by which bacteria reduce atmospheric nitrogen into ammonia is _____ _____.

11. Some microbes fix nitrogen by acting in a relationship with plants called _____.

12. The nitrogen fixing enzyme that bacteria must have to fix nitrogen is _____.

13. Plants of the pea family that often enter a relationship with bacteria to fix nitrogen are called _____.

14. The process by which ammonia is oxidized to nitrites or nitrates is _____.

15. The process by which nitrates are reduced to nitrous oxide or nitrogen gas is _____.

16. A sulfur-containing gas with a rotten egg smell is _____.

17. Sulfate converted to hydrogen sulfide is called _____ _____.

18. Sulfur oxidizing bacteria can create environmental problems due to the formation of _____ _____ which _____ the pH.

19. A type of air-flow hood that suctions air away from the opening and filters it before expelling it is the _____ flow hood.

20. The organic nonliving components of soil is called _____.

21. The main human pathogens of soil are in the genus _____.

22. Swamp gas, the main carbon-containing product is _____.

23. BOD refers to _____ _____ _____.

24. When water contains large quantities of _____ _____, decomposers rapidly deplete the oxygen supply through oxidation.

25. The main producers of the ocean are _____.

26. Excessive growth of algae often follows nutrient enrichment of water called _____.

27. The presence of <u>Escherichia coli</u> in water indicates that it has been contaminated by _____ _____.

28. Testing for water involves three stages namely _____, _____, and _____.

29. The process of adding alum to water which acts to precipitate out suspended colloids is _____.

30. The inoculation of water samples into lactose broth tubes represents the _____ test.

31. When samples of positive lactose tubes are plated onto EMB plates, the test is called _____.

32. Used water with the wastes it contains is called _____.

33. In the treatment of sewage, the use of physical means to remove solid wastes from sewage is _____ treatment.

34. The use of biological means to remove residual solid wastes is _____ treatment.

35. Chemical and physical means used to produce an effluent pure enough to drink is _____ treatment.

36. Spreading sewage over a bed of rocks is a treatment system called _____ _____.

37. Sludge that floats to the surface of water in the treatment process is known as _____.

38. An anaerobic chamber in which sludge from both primary and secondary treatments is digested is called a _____ _____.

39. Water that is fit for human consumption is termed _____ water.

Match the following terms with the description listed below.

Key Choices:

a. Nitrogen fixation   f. Sulfur oxidation
b. Nitrification   g. Hydrologic cycle
c. Denitrification   h. Carbon cycle
d. Sulfur reduction   i. Phosphorus cycle
e. Sulfate reduction

_____ 1. These bacteria are among the oldest life forms.

_____ 2. Conversion of carbon dioxide into organic molecules.

_____ 3. Conversion of ammonia to nitrates or nitrites.

_____ 4. Conversion of nitrates to nitrites and to nitrous oxide.

_____ 5. Oxidation of sulfur to sulfate and to sulfuric acid.

_____ 6. Requires a nitrogenase.

_____ 7. Converts sulfates to hydrogen sulfide.

_____ 8. Represents a wasteful process due to conversion of nitrates.

_____ 9. Involves relationships with plants called legumes.

_____ 10. Converts organic phosphates to inorganic forms.

_____ 11. Often responsible for increasing the acidity of streams due to oxidation of sulfide coal-mining wastes.

Match the following terms with the description listed below.

Key Choices:

a. Clostridium   e. Escherichia coli
b. Nitrosomonas   f. Methanococcus
c. Rhizobium   g. Sphaerotitus
d. Desulfovibrio

_____ 12. A sheathed bacterium which can clog sewage treatment systems.

_____ 13. Produce nitrates from ammonia.

_____ 14. Reduce sulfates to hydrogen sulfide.

_____ 15. Found in waterlogged soils of swamps; produces methane gas.

_____ 16. Commonly associated with legumes to fix atmospheric nitrogen.

_____ 17. Anaerobic spore-forming rods; soil pathogens.

_____ 18. Indicator organism of human fecal contamination.

REVIEW QUESTIONS

True/False (Mark T for True, F for False)

_____ 1. Denitrification is a very useful process because it helps to remove nitrates which are extremely toxic to plants.

_____ 2. Although microorganisms do not grow in air, spores of others can be transported on dust particles and water droplets.

_____ 3. As water becomes warmer it can support an increasing amount of oxygen and hence a greater variety of microbes.

_____ 4. The presence of coliforms generally indicates fecal pollution.

_____ 5. Following secondary treatment of sewage, the resultant water is cleaned of all organic material and is safe to drink.

_____ 6. Sulfate-reducing bacteria are among the oldest life forms.

Multiple Choice

_____ 7.   The steps necessary to purify water are (in order):
   a.   chlorination, flocculation, filtration, settling
   b.   flocculation, chlorination, filtration, settling
   c.   settling, flocculation, filtration, chlorination
   d.   settling, filtration, flocculation, chlorination

_____ 8.   Select the most correct statement.
   a.   Compared to fresh water, the ocean is much more variable in temperature and pH.
   b.   Marine organisms must tolerate varying degrees of salinity which changes with the depth.
   c.   Nutrient concentrations vary in ocean water often depending on depth and proximity to the shore.
   d.   Terrestrial and fresh water environments occupy more of the earth's surface than marine environments.

_____ 9.   Water is considered polluted when:
   a.   the BOD is very low.
   b.   a substance or condition is present that renders the water useless for a particular purpose.
   c.   the producers outnumber the consumers.
   d.   heterotrophic microbes are present.
   e.   all of the above are true.

_____ 10. In terms of water pollution all the following are true except:
   a.   heat can act as an important water pollutant.
   b.   excessive plant nutrients could lead to overgrowth of undesirable algae.
   c.   sediments could deplete the oxygen content of water and reduce visibility.
   d.   inorganic chemicals could alter the acidity of water.
   e.   all of the above are true.

ANSWER KEY

Fill-in-the-Blank

1. Ecology   2. Indigenous   3. Producers   4. Decomposers
5. Biogeochemical   6. Hydrologic   7. Greenhouse effect   8. CO_2
9. N_2 10. Nitrogen fixation   11. Symbiosis   12. Nitrogenase
13. Legumes   14. Nitrification   15. Denitrification   16. H_2S
17. Sulfate reduction   18. Sulfuric acid, lowers   19. Laminar
20. Humus   21. Clostridium   22. CH_4 23. Biological oxygen demand
24. Organic matter   25. Phytoplankton   26. Eutrophication
27. Fecal matter   28. Presumptive, confirmed, completed
29. Flocculation 30. Presumptive   31. Completed   32. Sewage
33. Primary   34. Secondary   35. Tertiary   36. Trickling filter
37. Bulking   38. Sludge digester   39. Potable

Matching

1.e 2.h 3.b 4.c 5.f 6.a 7.e 8.c 9.a 10.i 11.f

12.g 13.b 14.d 15.f 16.c 17.a 18.e

Review Answers

1.   **False**   Nitrates are very useful for plant growth and their removal would interfere with their metabolism. (p. 726)

2.   **True**   Endospores of bacteria, pollen of plants, and asexual spores of molds are commonly carried by wind currents, dust, and water droplets in the air. During dust storms and heavy pollution the microbial count can be very high. (p. 729)

3.   **False**   Moderate temperatures can support a wide variety of microbes. As water becomes very warm (above 50°C) it is more difficult to support useful organisms and especially fish because oxygen will begin to come out of solution. (pp. 733-734)

4.   **True**   Coliforms including <u>Escherichia</u> <u>coli</u> are those Gram-negative, non-spore forming aerobic or facultative anaerobic bacteria that ferment lactose and produce acid and gas. They are commonly associated with the GI tract. (pp. 736-737)

5.   **False**   The effluent from secondary treatment contains only 5 to 20% of the original quantity of organic matter. It can be discharged into streams but can cause problems and is not safe to drink. (p. 739)

6.   **True**   Sulfate reducers may be up to 3 billion years old and are found in anaerobic environments. They reduce sulfate to hydrogen sulfide. (pp. 727-728; Figure 26.8)

7.   c   Water is first settled to remove large suspended particles; then alum is added to cause flocculation; then it's filtered in a bed of sand; and finally it's chlorinated. (p. 737)

8.   c   Ocean environments are very consistent in temperature and pH; salinity of ocean waters does not vary with depth; ocean waters cover more territory than all the terrestrial and fresh water environments combined. (pp. 733-734)

9.   b   Water pollution is a relative one depending on the nature of the pollutants and the intended uses of the water. When the BOD is high the water can be depleted of oxygen rapidly; pollution is not related to a ratio of producers or consumers; pollution is not affected by the presence of heterotrophs. (pp. 735-736)

10.   e   Water pollution can be affected by a wide variety of factors, especially the ones listed here. (pp. 735-736)

CHAPTER 27

APPLIED MICROBIOLOGY

Every type of food that people might eat or drink is food that microbes love to eat as well. That has to be a rather chilling thought. This chapter will dramatically illustrate the fact that all kinds of foods can harbor and eventually be spoiled by an amazing variety of microbes which includes bacteria, yeasts, molds, algae, and even protozoa! Most of these microbes are harmless and only make the food products a bit less palatable but, unfortunately, some of these organisms are capable of producing poisonous toxins or multiplying to such an extent that they can cause disease. As a result, numerous attempts have been made to control these agents in the foods we eat, some more successful than others.

Interestingly, microbes have also been served as the food themselves. To date, the main attempts have been focused on their use as animal feed but the potential for human use seems to be quite good. Maybe all we need is an effective marketing tool.

The last part of this chapter is devoted to the use of microbes in industry. For centuries, microbes have been used to provide special flavors and to ferment our beverages. Today their uses have been expanded to the production of organic chemicals and pharmaceuticals, to waste disposal, and even to their use in mining. One certainly wonders what we would have done without these amazing microbes!

STUDY OUTLINE

I. **Microorganisms Found in Food**
   A. Grains
   B. Fruits and vegetables
   C. Meats, poultry, fish, and shellfish
   D. Milk
   E. Other edible substances
II. **Diseases Transmitted in Food**
   A. Effect of toxins
   B. Bacterial pathogens transmitted through food
   C. Viral agents transmitted through food
   D. Microbes transmitted through milk
III. **Prevention of spoilage and Disease Transmission**
   A. Food preservation
   B. Pasteurization of milk
   C. Standards regarding food and milk production
IV. **Microorganisms as Food and in Food Production**
   A. Algae and fungi as food
   B. Food production
V. **Industrial and Pharmaceutical Microbiology**
   A. Genetic engineering
   B. Metabolic processes applicable in industry
   C. Problems of industrial microbiology
   D. Beer, wine, and spirits
   E. Useful organic products
VI. **Microbiological Mining**
   A. Copper mining
   B. Other mineral mining
VII. **Microbiological Waste Disposal**
   A. Basic problems
   B. Degradation of toxins

REVIEW NOTES

A. **What kinds of microorganisms are found in different categories of food?**

Anything that people eat or drink is food for microorganisms. Since foods are derived from plants or animals which are associated with the soil, they will automatically have soil organisms on them. Fortunately most of these are not pathogenic but they can cause food spoilage. As long as the food is properly prepared, handled, and stored there is little chance for illness or spoilage.

Since harvested grains are normally dry there is little chance for microbial contamination. If they become moist because of improper storage, contamination especially by molds can occur. Flour, however, is purposely inoculated with yeasts to make bread.

Fruits and vegetables are usually protected by their resistant skins; however, if they are not properly stored, they will succumb to soft rot and mold damage.

Meats, poultry, fish, and seafood contain many kinds of microorganisms, some of which cause zoonoses. Because they are neutral in pH, they must be handled and prepared very carefully to keep dangerous pathogens from developing. Poultry often is contaminated with Salmonella, Clostridium perfringens, and Staphylococcus aureus. Fish and shellfish can be contaminated with several kinds of bacteria such as Vibrio's and viruses such as hepatitis A.

Milk can contain organisms from the cows, milk handlers, and the environment. Unfortunately, many of these bacteria can be potentially dangerous and, as a result, have required the pasteurization of milk and milk products.

Other substances such as sugar can support the growth of microbes but they are normally killed during the refining process. Spices are often used to mask the effects of microbes in foods; however, they can be the source of these microbes. See Table 27.1 which illustrates the numbers of bacteria in spices. Syrups for carbonated beverages can be contaminated with molds, and even tea, coffee, and cocoa can be subject to these molds if not kept dry.

B.   How are diseases transmitted in food?

Numerous types of diseases such as shigellosis, salmonellosis, botulism, and various food poisonings can be transmitted in food and milk. **Table 27.2** lists the most important organisms, diseases, and vectors.

TABLE 27.2   Pathogenic organisms transmitted in food and milk

Organism	Disease	Vector
Staphylococcus aureus	Food poisoning	Infected food handlers, unrefrigerated foods, milk from infected cows
Clostridium perfringens	Food poisoning	Unrefrigerated foods
Bacillus cereus	Food poisoning	Unrefrigerated foods
Clostridium botulinum	Botulism	Inadequately processed canned goods
Salmonella species	Salmonellosis	Infected food handlers, poor sanitation, seafood from contaminated water
Shigella species	Shigellosis	Infected food handlers, poor sanitation
Enteropathogenic *Escherichia coli*	"Montezuma's revenge" and other diseases	Infected food handlers (sometimes asymptomatic), poor sanitation
Campylobacter	Gastroenteritis	Undercooked chicken and raw milk
Vibrio cholerae	Cholera	Poor sanitation
Vibrio parahaemolyticus	Asian food poisoning	Undercooked fish and shellfish

---

Good sanitation and proper food handling, preparation, and storage greatly reduce the chances of getting food-borne diseases.

C.   How can food spoilage and disease transmission be prevented, and what standards relate to these problems?

The most important factor in preventing spoilage and disease transmission in food and milk is cleanliness in handling. Other factors include common-sense rules such as proper refrigeration, prompt use of fresh foods, and proper processing and storage.

Methods of food preservation include canning, refrigeration, freezing, lyophilization, drying, ionizing radiation, and the use of chemical additives such as sulfur dioxide, salt, and nitrates and nitrites. Most of these have been described in Chapter 10. Canning is the most common method of food preservation and involves the use of moist heat under pressure. If properly done, canning destroys all harmful spoilage microbes, prevents spoilage, and avoids any hazards of disease transmission.

Milk and milk products are rendered safe for human consumption by pasteurization or sterilization. This is done to avoid diseases such as tuberculosis, salmonellosis, brucellosis, and Q-fever.

Certain standards regarding food and milk production are maintained by federal, state, and local laws. Many states have banned the sale of raw milk to avoid the threat of milk-borne diseases. Tests for milk quality are summarized in **Table 27.3**.

TABLE 27.3   Tests used to determine milk quality

Test	Description	Purpose and Significance
Phosphatase test	Detects the presence of phosphatase, an enzyme-destroyed during pasteurization.	To determine whether adequate heat was used during pasteurization. If active phosphatase remains, pathogens also might be present.
Reductase test	Indirectly measures the number of bacteria in milk. The rate at which methylene blue is reduced to its colorless form is directly proportional to the number of bacteria in a milk sample.	To estimate the number of bacteria in a milk sample. High-quality milk contains so few bacteria that a standard concentration of methylene blue will not be reduced in $5\frac{1}{2}$ hours. Low-quality milk has so many that methylene blue is reduced in 2 hours or less.
Standard plate count	Directly measures viable bacteria. Diluted milk is mixed with nutrient agar and incubated 48 hours; colonies are counted, and number of bacteria in original sample is calculated.	To determine the number of bacteria in a milk sample. The number per milliliter must not exceed 100,000 in raw milk before pooled with other milk or 20,000 after pasteurization.
Test for coliforms	Same as the test used for water. (See Chapter 26.)	To determine whether coliforms are present. A positive coliform test indicates contamination with fecal material.
Test for pathogens	Detect the presence of pathogens. Methods vary depending on the suspected pathogens.	To identify pathogens. Usually not required but can help to locate the source of infectious agents that may appear in milk.

D.   How are microorganisms used as food or in the making of food products?

The rapid rise in world population is greatly increasing the demand for new and inexpensive sources of human food. Various microorganisms, especially yeasts, have been developed to provide some of the proteins and vitamins that would be required. Problems associated with this development include the expensive equipment and the difficulty in getting people to accept the food as part of their diet. Algae has also been used as a protein food supplement and can be readily grown in lakes and on sewage. Problems include the danger of viral contamination and the lack of its acceptability as a food.

Yeasts are commonly used in food production to leaven bread and to ferment wines and beer. Certain bacteria are used to make dairy products such as buttermilk, sour cream, yogurt, and cheeses. In cheese-making, the whey of milk is discarded and microorganisms ferment the curd and impart flavors and texture to the product. Table 27.5 provides a classification of ripened cheeses, examples, and the organisms associated with the ripening.

Many other foods are produced by microbial action and include vinegar, sauerkraut, pickles, olives, poi, soy sauce, and other soy products. At the next opportunity, go to the grocery store and make a list of all the foods that are affected in some way by microbial action. The list will be quite lengthy.

E.   How can microbes be used in industry, and what problems are associated with their use?

Industrial microbiology deals with the use of microorganisms to assist in the manufacture of useful products such as simple organic compounds and alcoholic beverages, or to dispose of waste products such as garbage and waste paper. Pharmaceutical microbiology deals with the use of microorganisms in the manufacture of medically useful products such as antibiotics, vitamins, and hormones.

The production of complex molecules and metabolic end products in commercially profitable quantities usually requires the manipulation of microbial processes. These manipulations include altering the nutrients, or the environmental conditions, and isolating mutants and/or modifying them by genetic engineering.

Problems include adapting small-scale processes into large-scale commercial processes and developing techniques that are necessary for recovery of the products.

F.   How are microbes used in the manufacture of beer, wines, and spirits?

Beer and wine are made by fermenting sugary juices; spirits, such as whiskey, gin, and rum, are made by fermenting juices and distilling the fermented product. Strains of Saccharomyces yeasts are used as the fermenters for all alcoholic beverages. The yeasts produce ethyl alcohol, carbon dioxide, and other substances such as acetic and butyric acids and tannin which add much to the flavor of the beverage.

G.   What is the role of microorganisms in the manufacture of simple organic compounds, antibiotics, enzymes, and other biologically useful substances?

Microorganisms are capable of producing a wide variety of useful organic compounds including alcohols, acetone, glycerol, and organic acids. Currently, microbes play a limited industrial role; however, they may become more important in the future because of advances in genetic engineering.

Microbes such as Streptomyces, Penicillium, Cephalosporin, and Bacillus are greatly involved in producing a wide variety of antibiotics. New strains are continuously being discovered that are capable of producing potential substances. Some antibiotics such as the penicillins have beta-lactam rings that can be modified in the laboratory to produce more effective agents.

Various enzymes including proteases and amylases have been extracted from microorganisms such as Aspergillus and Bacillus and are produced commercially. Many of these are used in detergents, drain cleaners, enrichment agents of foods, and in the manufacture of paper.

Vitamins and hormones are now made by manipulating organisms so that they produce excessive amounts of the products. Single-cell proteins consist of whole organisms such as Candida that are grown in simple foods to produce large quantities of protein. They are used primarily as animal feed.

H.   How are microorganisms used in mining?

Since the availability of mineral-rich ores has decreased, various methods must be used to extract the minerals from less concentrated sources. Microbes have been used to extract copper from low-grade ores as well as small quantities of iron, uranium, arsenic, lead, zinc, cobalt, and nickel.

I.   How are microorganisms used in waste disposal?

Microbes are very important components in sewage treatment plants as noted in Chapter 26. A few organisms have been found to degrade toxic wastes and certain types of plastics. Microbes have also been used to degrade oil products lost during spills.

Research is under way to identify and develop other uses for microbes.

LEARNING ACTIVITIES

Complete each of the following items by supplying the appropriate word or phrase.

1. The yeast commonly used to make bread dough rise is _____.

2. The most common bread mold is probably the black spored mold _____ _____.

3. The microbe that caused large numbers of Irish to immigrate to this country in 1846 was a _____.

4. A parasitic worm that is found in raw pork samples is _____.

5. Salmonella responsible for many outbreaks of food poisonings are commonly associated with animal reservoirs especially _____.

6. Shellfish such as oysters and clams often carry a Gram-negative curved rod called _____ that causes serious intestinal infections.

7. When species of _Streptococci_ and _Lactobacilli_ act on milk, they usually result in _____.

8. A human disease caused by species of _Erysipelothrix_ that is contracted by eating infected pork is _____.

9. Sugars and jellies act to preserve foods by _____ _____.

10. The normal refrigeration temperature of _____ is enough to retard growth for only a few days.

11. Home freezers operating at about _____ are effective in preserving food for several months.

12. A method of preparing instant coffee is known as _____.

13. Ionizing radiation can sterilize foods but cannot destroy bacterial _____.

14. Organic acids such as benzoic and sorbic are added to foods to _____ the pH.

15. A gas that is used to control yeast in wines is _____ _____.

16. A chemical added to foods to alter the osmotic pressure could be _____ _____.

17. An anticlostridial agent that can be added to food is _____.

18. Milk that is heated to 71.6°C for 15 seconds is pasteurized by the _____ pasteurization process.

19. A milk test that detects the number of bacteria in milk based on the time it takes to reduce methylene blue is the _____ test.

20. Two types of microbes that could be used for human consumption are _____ and _____.

21. The microbe used to leaven bread is _____.

22. The genera commonly found in dairy products that imparts useful characteristics and flavors are _____, _____, and _____.

23. Coagulated milk proteins form into a cheesy mass called _____ and the liquid that remains is _____.

24. Vinegar contains the organic acid _____ acid.

25. A soft curd of soybeans is _____.

26. A food common in the South Pacific that is made from the fermented roots of the taro plant is _____.

27. Cabbage, if allowed to ferment by action of species of Lactobacillus and Leuconostoc, is _____.

28. An industrial system where fresh medium is introduced at one end and medium containing the product is withdrawn at the other is a _____ _____.

29. Strains of _____ are used in fermentation to produce all alcohol beverages.

30. In order to make beer, grains that are partially germinated or _____ are mixed with hot water and a plant called _____.

31. The main product of fermentation of grapes and grains is _____.

32. Fortified wines such as sherry and cognac have extra _____ added.

33. Enzymes added to detergents to increase their cleaning power are _____.

34. The flavor enhancer monosodium glutamate uses _____ acid which is derived from bacteria.

Match the following terms with the description listed below.

Key Choices:

a. <u>Saccharomyces cerevisiae</u>
b. <u>Pachysolen tannophilus</u>
c. <u>Penicillium chrysogenum</u>
d. <u>Aspergillus</u>
e. <u>Thiobacillus ferroxidans</u>
f. <u>Phytophthora infestans</u>
g. <u>Salmonella</u> species

h. <u>Streptococcus lactis</u>
i. <u>Erwinia dissolvens</u>
j. <u>Vibrio parahaemolyticus</u>
k. <u>Spirulina</u>
l. <u>Lactobacillus</u> species
m. <u>Escherichia coli</u>

_____ 1. Produces large quantities of alcohol from 5 carbon sugars.

_____ 2. Uses oxygen to oxidize sulfur atoms in sulfide ores.

_____ 3. Used to digest pectin in outer coverings of coffee beans.

_____ 4. Can be used to make huge quantities of citric acid and proteases.

_____ 5. Used to make acidophilus milk and yogurt.

_____ 6. Commonly causes sour milk.

_____ 7. Ferments grapes to produce wines.

_____ 8. Used in making bread.

_____ 9. Commonly found with poultry and eggs; capable of causing food poisoning.

_____ 10. Produces the antibiotic penicillin.

_____ 11. Causes food poisoning associated with raw oysters and undercooked shellfish.

_____ 12. Responsible for the Irish potato famine of 1846.

_____ 13. Can be transmitted by infected food handlers.

_____ 14. An alga grown in alkaline lakes; made into cakes and used as a food.

Match the following terms with the description listed below. Choices may be used more than once.

Key Choices:

a. Gravies
b. Fruits and vegetables
c. Meats
d. Poultry and eggs
e. Fish and shellfish
f. Milk
g. Chocolates
h. Hard candy
i. Spices

_____ 15. When kept dry, the product is rarely affected by microbes.

_____ 16. Can find commensals such as <u>Pseudomonas fluorescens</u>.

_____ 17. Salmonella are commonly found with this food.

_____ 18. <u>Rhizopus</u> and <u>Mucor</u> can produce fluffy, white growth on the surface.

_____ 19. Can be associated with trichinosis.

_____ 20. Can be infected with <u>Mycobacteria</u> and <u>Brucella</u>.

_____ 21. Large concentrations of vibrios can be found.

_____ 22. Often used to mask odors and can carry large numbers of commensals.

_____ 23. Generally safe from microbial contamination.

REVIEW QUESTIONS

True/False (Mark T for True, F for False)

_____ 1. The skins of most plant foods are very inhibitory to microbial agents.

_____ 2. Since eggs have very hard shells, they are rarely associated with any microbial agents and are not considered sources of infections.

_____ 3. Foods that contain artificial sweeteners do not retard growth by osmotic pressures and therefore required more elaborate methods to inhibit microbes.

_____ 4. The manufacture of vinegar requires the action of vinegar eels to produce the acid.

_____ 5. Wines can be made from any fruit.

_____ 6. By adding more sugar to a wine, the alcohol content can be increased to as much as 40%.

_____ 7. All enzymes used in industrial processes are synthesized by living organisms.

_____ 8. Flat sour spoilage of canned foods occurs when bacterial endospores germinate, grow, and spoil the food but do not cause the cans to bulge with gas.

Multiple Choice

_____ 9. The use of nitrates and nitrites in food:
a. should be stopped immediately because when heated, they are converted to formaldehyde.
b. are effective in controlling diseases such as botulism in sausages.
c. cause meats to turn a dull brown color.
d. retard the maturation of fruits and resist spoilage.

_____ 10. Foods that are irradiated:
a. become slightly radioactive.
b. can be sterilized.
c. can be stored at room temperature indefinitely.
d. will cause considerable color changes in the product.

_____ 11. Which of the following is not true concerning the addition of antibiotics to animals used for food.
a. Humans may become allergic to antibiotics given to animals.
b. The antibiotics might interfere with the activities of microbes such as those essential in fermenting milk and cheese.
c. The antibiotics might be relied on instead of good sanitation.
d. The antibiotics might reduce the number of resistant strains of bacteria thus making the foods safer to eat.
e. All of the above are true.

_____ 12. In the production of cheeses:
a. hard cheeses ripen in about two weeks.
b. Species of Clostridium are commonly used to produce the best flavors in cheddar and Roquefort.
c. Hard cheeses are often very large and are ripened by microbial action.
d. Soft cheeses are ripened by the action of salt.

ANSWER KEY

Fill-in-the-Blank

1. <u>Saccharomyces</u> 2. <u>Rhizopus nigricans</u> 3. Fungus 4. Trichinosis
5. Poultry 6. Vibrio 7. Souring 8. Erysipeloid
9. Osmotic pressures 10. 4°C 11. -10°C 12. Lyophilization
13. Enzymes 14. Lower 15. Sulfur dioxide 16. Sodium chloride
17. Nisin 18. Flash 19. Reductase 20. Yeast, algae 21. Yeast
22. <u>Streptococcus</u>, <u>Leuconostoc</u>, <u>Lactobacillus</u> 23. Curd, whey
24. Acetic 25. Tofu 26. Poi 27. Sauerkraut
28. Continuous reactor 29. Saccharomyces 30. Malted, hops
31. Alcohol 32. Alcohol 33. Proteases 34. Glutamic

Matching

1.b 2.e.3.i 4.d 5.1 6.h 7.a 8.a 9.g 10.c 11.j 12.f 13.m 14.k

15.a-i 16.b 17.d 18.c 19.c 20.f 21.e 22.i 23.h

Review Answers

1. **True** Plant skins are often very tough and contain waxes and may even release antibiotic substances to inhibit microbial invasion. (p. 750)

2. **False** Most eggs are free of contamination; however, some bacteria such as pseudomonads and some fungi can grow on eggshells. <u>Salmonella</u> can also survive on eggshells and can be involved in disease transmission. (pp. 751-753)

3. **True** Artificial sweeteners can not retard growth by osmotic effects as foods containing natural sugars. High sugar content can create a hypertonic condition. (pp. 757-758)

4. **False** Vinegar is made from ethyl alcohol by the action of acetic acid bacteria and not eels. (p. 765; Figure 27.9)

5. **True** Wine is made from juice extracted usually from grapes although any fruit including nuts and certain blossoms can be used. Normally, sulfur dioxide is used to kill any wild yeasts that may lurk on any of the fruits being used for fermentation. (p. 768)

6. **False** Once the alcohol content reaches 12 to 15 percent, it poisons the yeasts carrying out the fermentation. Extra alcohol is added after the wine is made in order to increase the alcohol content. (p. 768)

7.   **True** Enzymes are made of protein and are produced by protein-synthetic mechanisms found in living organisms. A few enzymes are extracted from plants but most are produced by bacteria.
(p. 770; Figure 27.17)

8.   **True** When spores germinate and produce gas, the resultant condition is known as thermophilic anaerobic spoilage. (p. 757)

9.   **b** When heated, nitrates and nitrites may be converted to carcinogenic nitrosamines; nitrates and nitrites cause meats to remain bright, fresh, and red looking; carbon dioxide retards maturation of fruits. (p. 759)

10.   **b** Radiation does not affect foods and they do not become radioactive. Radiation does not destroy enzymes; therefore, the food, if left at room temperature, may be affected by autodigestion. Radiation causes no color or odor changes. (p. 758)

11.   **d** Antibiotics should not be used to reduce the number of bacteria in foods because of the threat of resistant strains developing. (p. 759)

12.   **c** Hard cheeses require 2 to 16 months to ripen; Clostridium species are soil anaerobes that would produce terrible flavors in any milk product; salt will actually remove moisture from cheeses to create a firmer product. (pp. 763-764)

APPENDIX

ADDITIONAL READINGS

Chapter 1:   Scope and History of Microbiology

Baker, J.J.W. 1968. *Hypothesis, prediction, and implication in biology*. Reading, Mass.: Addison-Wesley.

Dubos, R. 1974. Pasteur's dilemma - the road not taken. *American Society of Microbiology News* 40:703.

Jawetz, E., J.L. Melnick, and E.A. Adelberg. 1991. *Review of medical microbiology*. East Norwalk, Ct: Appleton & Lange.

Chapter 2:   Fundamentals of Chemistry

Beebe, G.W. 1982. Ionizing radiation and health. *American Scientist* 70 (January-February):35

Bohinski, R.C. 1987. *Modern concepts of biochemistry,* 3rd ed. Boston: Allyn and Bacon.

Gibbons, A. and M. Hoffman. 1991. New 3-D protein structures revealed. *Science* 253 (July 26):382-83.

Inoue, M. and M. Koyano. 1991. Fungal contamination of oil paintings in Japan. *Int. Biodeterior.* 28 (1-4):23-35.

Upton, A.C. 1982. The biological effects of low-level ionizing radiation. *Scientific American* 246, no.2 (February):41.

Wiggins, P.M. 1990. Role of water in some biological processes. *Microbiological reviews* 54 (4):432-49.

Chapter 3:   Microscopy and Staining

Anonymous. 1990. Unique atomic views from STM's new kin. *Science News* 137, no.14 (April 7):223.

Arscott, P.G. 1990. Taking the measure of the molecule by scanning tunneling microscopy. *ASM News* 56 (3):136-38.

England, B.M. 1991. The state of the science: scanning electron microscope. *Mineralogical Record* 22, no.2 (March-April): 123-33.

Howells, M.R., J. Kirz, and D. Savre. 1991. X-ray microscopes. *Scientific American* 264 (February):88-94.

Molina, T.C., H.D. Brown, and R.M. Irbe. 1990. Gram staining apparatus for space station applications. *Applied and Environmental Microbiology* 56 (March):601-606.

Tang, S.L. 1991. Scanning tunneling microscopy. *Chemtech* 21, no.3 (March):182-90.

Trifiro, S., A. Bourgault, F. Lebel, and P. Rene. 1990. Ghost mycobacteria on gram stain. *Journal of Clinical Microbiology* 28 (1):146-47.

Trux, J. 1991. Through the looking glass. *World Magazine* 58, no.8 (March):58-64.

Woeste, S. and P. Demchick. 1991. New version of the negative stain. *Applied and Environmental Microbiology* 57 (6):1858-59.

Chapter 4: Characteristics of Prokaryotic and Eukaryotic Cells

Hancock, R.E.W. 1991. Bacterial outer membranes: evolving concepts. *ASM News* 57 (4):175-82.

Koch, A.L. 1988. Biophysics of bacterial walls viewed as stress-bearing fabric. *Microbiological Reviews* 52 (3):337-53.

Koch, A.L. 1990. Growth and form of the bacteria growth. *American Scientist* 78, no.4 (July-August):327-42.

Koppel, T. 1991. Learning how bacteria swim could set new gears in motion. *Scientific American* 265 (September):168-69.

Shapiro, J.A. 1991. Multicellular behavior of bacteria. *ASM News* 57 (5):247-53.

Chapter 5: Essential Concepts of Metabolism

Amesz, V.J. 1988. Primary processes in bacterial photosynthesis. *Biol. Rundsch* 26 (4):185-95.

Meighen, E.A. 1991. Molecular biology of bacterial bioluminescence. *Microbiological Reviews* 55 (1):123-42.

Pritchard, P.H. 1991. Bioremediation as a technology: experiences with the Exxon Valdez oil spill. *Journal of Hazardous Material* 28 (1-2):115-30.

Trumpower, B.L. 1990. Cytochrome bc complexes of microorganisms. *Microbiological Reviews* 54 (2):101-29.

Chapter 6: Growth and Culturing of Bacteria

Anonymous. 1991. Life in hell. *Discover* 12 (February):12.

Anonymous. 1991. Salty life on Mars. *Discover* 12 (June):12.

Barker, L.P., W.A. Simpson, and G.D. Christensen. 1990. Differential production of slime under aerobic and anaerobic conditions. *Journal of Clinical Microbiology* 28 (11):2578-79.

Berger, S.A., B. Bogokowsky, and C. Block. 1990. Rapid screening of urine for bacteria and cells by using a catalase reagent. *Journal of Clinical Microbiology* 28 (5):1066-67.

Foster, J.W. 1992. Beyond pH homeostasis: the acid tolerance response of Salmonellae. *ASM News* 58 (5):266-70.

Herring, T.S. 1990. Microbiologists gear up to encourage better clinical laboratory standards; new federal rules for proficiency testing, personnel, and waived tests worry scientists in clinical laboratories. *Scientist* 4 (10):2.

Pasarell, L. and M.R. McGinnis. 1992. Viability of fungal cultures maintained at -70°C. *Journal of Clinical Microbiology* 30 (4):1000-1004.

Waters, J.R. 1992. Detection of bacterial growth by gas absorption. *Journal of Clinical Microbiology* 30 (5):1205-1209.

Chapter 7:   Genetics I: Gene Action, Gene Regulation, and Mutation

Anonymous. 1990. Bacterium bags UV light. *BioScience* 40 (June):479.

Bachmann, B.J. 1990. Linkage map of <u>Escherichia</u> <u>coli</u> K-12 Edition 8. *Microbiological Reviews* 54 (2):130-97.

Cairns, J. 1986. The myth of the most sensitive species. *BioScience* 36 (November):670.

Howard-Flanders, P. 1981. Inducible repair of DNA. *Scientific American* 245, no.5 (November):72.

Kazazian, H.H. 1985. The nature of mutation. *Hospital Practice* 20:55.

Chapter 8:   Genetics II: Transfer of Genetic Material and Genetic Engineering

Ahern, H. 1991. Self-splicing introns: molecular fossils or selfish DNA? *ASM News* 57 (5):258-61.

Anderson, S.G.E. and C.G. Kurland. 1990. Codon preferences in free-living microorganisms. *Microbiological Reviews* 54 (2):198-210.

Anonymous. 1990. Questions also linger over agricultural biotechnology. *ASM News* 56 (12):631-32.

Barton, J.H. 1991. Patenting life. *Scientific American* 264 (March):40-46.

Couturier, M., F. Bex, P.L. Bergquist, and W.K. Maas. 1988. Identification and classification of bacterial plasmids. *Microbiological Reviews* 52 (3):375-95.

Fox, J.L. 1990. Engineered rats reveal arthritic surprise. *Science News* 138 (December 8):357.

McKnight, S.L. 1991. Molecular zippers in gene regulation. *Scientific American* 264 (April):54-58.

Rennie, J. 1991. Proofreading genes: a molecular editor makes sensible additions to RNA. *Scientific American* 264 (May):28.

Moss, R. 1991. Genetic transformation of bacteria. *The American Biology Teacher* 53 (March):179-80.

Newman, S.A. 1990. The difficulties of patenting transgenic animals. *ASM News* 56 (5):252-53.

Thomson, R.G. 1989. Recombinant DNA made easy: gene, gene, who's got the gene? *American Biology Teacher* 54, no.4 (April): 226-233.

Van Houten, B. 1990. Nucleotide excision repair in <u>Escherichia</u> <u>coli</u>. *Microbiological Reviews* 54 (1):18-51.

Chapter 9:   Microbes in the Scheme of Life: An Introduction to Taxonomy

Gillis, A.M. 1991. Can organisms direct their evolution? *BioScience* 41 (April):202-06.

Hammer, M., A. Coghlan, and T. Toro. 1990. Will taxonomy go the way of the dinosaurs? *New Scientist* 126 (June 23):32-33.

Schopy, J.W. 1978. The evolution of the earliest cells. *Scientific American* 239 (September):110-12.

Vidal, G. 1984. The oldest eukaryotic cells. *Scientific American* 250, no.2 (February):48.

Zillig, W. 1987. Eukaryotic traits in archaebacteria. *Annals of the New York Academy of Sciences* 503:7.

Chapter 10: The Bacteria

Anonymous. 1986. Missing mycoplasmas. *Science News* 130 (September 20):104.

Belas, R. 1992. The swarming phenomenon of <u>Proteus</u> <u>mirabilis</u>. *ASM News* 58 (1):15-22.

Childress, J.J., H. Felbeck, and G.N. Somers. 1987. Symbiosis in the deep sea. *Scientific American* 256 (May):114-20.

Ferris, M.J. and C.F. Hirsch. 1991. Method for isolation and purification of cyanobacteria. *Applied and Environmental Microbiology* 57 (May):1448-52.

Fischetti, V.A. 1991. Streptococcal M protein. *Scientific American* 264 (June):58-65.

Harold, F.M. 1990. To shape a cell: an inquiry into the causes of morphogenesis of microorganisms. *Microbiological Reviews* 54 (4):381-431.

Shapiro, J.A. 1988. Bacteria as multicellular organisms. *Scientific American* 258, no.6 (June):82.

Shimkets, L.J. 1990. Social and developmental biology of the myxobacteria. *Microbiological Reviews* 54 (4):473-501.

Chapter 11: Viruses

Anonymous. 1990. Cats and AIDS. *Discover* 11 (July):16.

Anonymous. 1992. Prions revisited - no virus, but conformational changes in proteins. *ASM News* 58 (5):256-57.

Anonymous. 1991. Virus sculptures. *Discover* 12 (June):10-12.

Barlow, R.M. and D.J. Middleton. 1990. Dietary transmission of bovine spongiform encephalopathy to mice. *Vet Rec* 126:111-112.

Beardsley, T. 1990. Tainted feed, mad cows: could a British cattle disease infect U.S. herds? *Scientific American* 262 (May):34.

Caldwell, M. 1991. Mad cows and wild proteins. *Discover* 12 (April):68-72.

Dajer, T. 1990. A herpes key. *Discover* 11 (November):22.

Gibbs, C.J., J. Safar, M. Ceroni, M.A. Di, W.W. Clark, and J.L. Hourrigan. 1990. Experimental transmission of scrapie to cattle. *Lancet* 335:1275.

Lai, M.M.C. 1992. RNA recombination in animal and plant viruses. *Microbiological Reviews* 56 (1):61-79.

Montgomery, G. 1990. The ultimate medicine. *Discover* 11 (March): 60-67.

Pringle, C.R. 1992. Virus "species" now defined. *ASM News* 58 (2):70-71.

Prusiner, S.B. 1991. Molecular biology of prion diseases. *Science* 252, no.5012 (June 14):1515-23.

Roossinck, M.J., D. Sleat, and P. Palukaitis. 1992. Satellite RNAs of plant viruses: structure and biological effects. *Microbiological Reviews* 56 (2):265-279.

Webster, R.G., W.J. Bean, O.T. Gorman, T.M. Chambers, and Y. Kawaoka. 1992. Evolution and ecology of influenza A viruses. *Microbiological Reviews* 56 (1):152-179.

Chapter 12:   Eukaryotic Microorganisms and Parasites

Anonymous. 1991. Fungi as human pathogens: sublime and bizarre. *ASM News* 57(1):7.

Brieger, W.R., J. Ramakrishna, J.D. Adeniyi, M.K.C. Sridhar, and O.O. Kale. 1991. Guinea worm control case study: planning a multi-strategy approach. *Social Science & Medicine* 32, no.12 (June):1319-28.

Cherfas, J. 1991. Disappearing mushrooms: another mass extinction? *Science* 254, no.5037 (December 6):1458.

Cooter, R.D., I.S. Lim, D.H. Ellis, and I.O.W. Leitch. 1990. Burn wound Zygomycosis caused by Apophysomyces elegans. *Journal of Clinical Microbiology* 28 (9):2151-53.

Eissenberg, L.G. and W.E. Goldman. 1991. Histoplasma variation and adaptive strategies for parasitism: new perspectives on histoplasmosis. *Clinical Microbiology Reviews* 4 (4):411-21.

Hawksworth, D.L. 1990. Name changes in fungi of economic importance. *Mycopathologia* 111 (2):73-74.

Koske, R.E. and J.N. Gemma. 1990. Mycorrhizae in strand vegetation of Hawaii: evidence for long-distance codispersal of plants and fungi. *American Journal of Botany* 77, no.4 (April):466-75.

Maddison, S.E. 1991. Serodiagnosis of parasitic diseases. *Clinical Microbiology Reviews* 4 (4):457-69.

Orlowski, M. 1991. Mucor dimorphism. *Microbiological Reviews* 55 (2):234-58.

Sharnoff, S.K. 1991. Beauties from a beast: woodland Jekyll and Hydes. Anomalies among living things, slime molds are neither plants nor animals nor fungi, yet they hold their own among nature's jewels. *Smithsonian* 22, no.4 (July):98-104.

Chapter 13:   Sterilization and Disinfection

Balows, A., ed. 1991. *Manual of Clinical Microbiology*, 5th ed. Washington, D.C.: American Society for Microbiology.

Borick, P.M. and R.E. Peller. 1970. *Disinfection*. New York: Marcel Dekker.

Castle, M. 1980. *Hospital Infection Control*. New York: John Wiley and Sons.

Collins, C.H., P.M. Lyne, and J.M. Grange, eds. 1989. *Microbiological Methods*, 6th ed. Stoneham, Mass.: Butterworths.

Miller, B.M., ed. 1986. *Laboratory Safety: Principles and Practices*. Washington, D.C.: American Society for Microbiology.

Schmitz, A. 1988. Is this any way to treat your food? *Hippocrates*. (November-December).

Chapter 14: Antimicrobial Therapy

Alper, J. 1990. Computer-aided design of antiviral agents. *ASM News* 56 (5):269-71.

Anonymous. 1990. Antifungal agents. *Drug Store News* 12, no.14 (July 23):19-22.

Anonymous. 1990. Drugs for viral infections. *Medical Letter on Drugs and Therapeutics* 32, no.824 (August 10):73-79.

Brown, P. 1992. How drug defeats malaria parasite. *New Scientist* 133, no.1803 (January 11):21.

Cooper, M. 1990. Understanding viral replication is key to new therapeutic approaches. *CDC AIDS Weekly* (July 16):7-9.

DeNoon, D.J. 1990. Nucleoside analogs. *CDC AIDS Weekly* (January 8):36-40.

Fox, J.L. 1991. Case for DNA triple helix strengthens. *ASM News* 56 (4):202-206.

Merigan, T.C. 1991. Treatment of AIDS with combinations of antiretroviral agents. *American Journal of Medicine* 90, no.4A (April 10):85-95.

Strickland, D.A. 1990. Paired nucleoside analogs appear effective, tolerable. *Medical World News* 31, no.13 (July):15-17.

Tenover, F.C. 1991. Novel and emerging mechanisms of antimicrobial resistance in nosocomial pathogens. *American Journal of Medicine* 91 (3):76-81.

Wright, K. 1990. Bad new bacteria. *Science* 249, no.4964 (July 6):22-25.

Chapter 15: Host-Microbe Relationships and Disease Processes

Anonymous. 1991. Overall U.S. health shows some improvements. *ASM News* 57 (6):294-96.

Dowling, J.N., A.K. Saha, and R.H. Glew. 1992. Virulence factors of the family <u>Legionellaceae</u>. *Microbiological Reviews* 56 (1): 32-60.

Kanatani, F.N. 1985. Use of the armadillo in Hansen's disease research. *Journal of Environmental Health* 47 (May-June): 310-11.

Luftig, R.B., D. Pieniazek, and N. Pieniazek. 1990. Update on viral pathogenesis. *ASM News* 56 (7):366-68.

Moulder, J.W. 1991. Interaction of chlamydiae and host cells in vitro. *Microbiological Reviews* 55 (1):143-90.

Romero-Steiner, S.T. Witek, and E. Balish. 1990. Adherence of skin bacteria to human epithelial cells. *Journal of Clinical Microbiology* 28 (1):27-31.

Szcepanski, A. and J.L. Benach. 1991. Lyme borreliosis: host responses to <u>Borrelia</u> <u>Burgdorferi</u>. *Microbiological Reviews* 55 (March):21-34.

Weber, A., K. Harris, S. Lohrke, L. Forney, and A.L. Smith. 1991. Inability to express fimbriae results in impaired ability of <u>Haemophilus influenzae</u> B to colonize the nasopharynx. *Infection and Immunity* 59 (12):4724-28.

Zwilling, B.S. 1992. Stress affects disease outcomes. *ASM News* 58 (1):23-25.

Chapter 16: Epidemiology and Nosocomial Infections

Anonymous. 1991. Elimination of nasal and hand carriage of <u>S. aureus</u>. *American Family Physician* 43 (April):1378.

Anonymous. 1991. Environmental contamination in child day-care centers. *The Journal of the American Medical Association* 266, no.15 (October 16):2071.

Anonymous. 1991. The risk of contracting HIV infection in the course of health care. *The Journal of the American Medical Association* 265, no.14 (April 10):1872-73.

Bernstein, B.J. 1987. Churchill's secret biological weapons. *Bulletin of the Atomic Scientists* 43 (January-February):46-51.

Decker, J.F. 1991. Depopulation of the Northern Plains natives. *Social Science & Medicine* 33, no.4 (August 15):381-94.

Jackson, M.M. and P. Lynch. 1990. In search of a rational approach. *American Journal of Nursing* 90, no.10 (October):65-71.

Joyce, C. 1988. Russians rebut claims of anthrax weapons disaster. *New Scientist* 118, no.1609 (April 21):24-25.

Kaiser, A.B. 1991. Surgical-wound infection. *New England Journal of Medicine* 324, no.2 (January 10):123-24.

Mathews, D.S., R.E. Pust, and D.H. Cordes. 1991. Prevention and treatment of travel-related illness. *American Family Physician* 44, no.4 (October):1343-59.

Moffat, A.S. 1990. China: a living lab for epidemiology. *Science* 248 (May 4):553-55.

Ross, V. 1990. Nosocomial infections during the age of AIDS. *ASM News* 56 (11):575-78.

Weinstein, R.A. 1991. Epidemiology and control of nosocomial infections in adult intensive care units. *The American Journal of Medicine* 91 (3):179-184.

Chapter 17: Host Systems and Nonspecific Host Defenses

Anonymous. 1989. Making interferon. *Chemtech* 19, no.7 (July): 428-30.

Baron, S., S.K. Tyring, W.R. Fleischchmann Jr., D.H. Coppenhaver, D.W. Niesel, G.R. Klimpel, J. Stanton, and T.K. Hughes. 1991. The interferons: mechanisms of action and clinical applications. *The Journal of the American Medical Association* 266, no.10 (September 11):1375-84.

Brooks, A.M. 1991. Inflammation information. *Current Health 2* 18, no.1 (September):16-19.

Cooper, E.L. 1990. Immune diversity throughout the animal kingdom. *BioScience* 40, no.10 (November):720-23.

Ealick, S.E., W.J. Cook, S. Vijay-Kumar, M. Carson, T.L. Nagabhushan, P.P. Trotta, and C.E. Bugg. 1991. Three-dimensional structure of recombinant human interferon-gamma. *Science* 252, no.5006 (May 3):698-702.

Peterson, H.B., C.K. Walker, J.G. Kahn, E.A. Washington, D.A. Eschenbach, and S. Faro. 1991. Pelvic inflammatory disease: key treatment issues and options. *The Journal of the American Medical Association* 266, no.18 (November 13):2605-12.

Kirby, P. 1988. Interferon and genital warts: much potential, modest progress. *The Journal of the American Medical Association* 259, no.4 (January 22):570-73.

Omata, M., O. Yokosuka, S. Takano, N. Kato, K. Hosoda, F. Imazeki, M. Tada, Y. Ito, and M. Ohto. 1991. Resolution of acute hepatitis C after therapy with natural beta interferon. *Lancet* 338, no.8772 (October 12):914-16.

Chapter 18: Immunology I: Basic Principles of Specific Immunity and Immunization

Anonymous. 1990. Vaccine-preventable diseases among adults: standards for adult immunization practice. *The Journal of the American Medical Association* 264, no.18 (November 14):2375-76.

Erickson, D. 1990. Of mice and men. How form affects function in monoclonal-antibody drugs. *Scientific American* 262 (April): 76-77.

Fox, J.L. 1990. Bounty of immunity. *ASM News* 56 (2):87-89.

Hoffman, M. 1990. "Superantigens" may shed light on immune puzzle. *Science* 248, no.4956 (May 11):685-86.

Powers, D.C., L.F. Fries, B.R. Murphy, B. Thumar, and M.L. Clements. 1991. In elderly persons, live attenuated influenza A virus vaccines do not offer an advantage over inactivated virus vaccine in inducing serum or secretory antibodies or local immunology memory. *Journal of Clinical Microbiology* 29 (3):498-505.

Rennie, J. 1990. Antibody bonanza. *Scientific American* 262 (February):62.

Rennie, J. 1990. Overview: tolerating self. *Scientific American* 263 (September):50.

Rennie, J. 1990. Who's the dealer? What controls gene shuffling in the immune system? *Scientific American* 262 (March):30.

Von Boehmer, H. and P. Kisielow. 1991. How the immune system learns about self. *Scientific American* 265 (October):74-81.

Zur Hausen, H. 1991. Viruses in human cancers. *Science* 254, no.5035 (November 22):1167-73.

Chapter 19: Immunology II: Immunologic Disorders and Immunologic Tests

Anonymous. 1991. Bronchial inflammation and asthma treatment. *Lancet* 337, no.8733 (January 12):82-84.

Anonymous. 1992. CDC to broaden key AIDS definitions. *ASM News* 58 (1):10

Anonymous. 1991. The first case. *Discover* 12 (January):74-76.
Anonymous. 1991. Unreported findings shed new light on HIV dental case. *AIDS Alert* (6):121-130.
Beardsly, T. 1991. Positive response: encouraging results in the search for an AIDS vaccine. *Scientific American* 265 (August):26.
Carpenter, C.C.J., K.H. Mayer, and M.D. Stein. 1991. Human immunodeficiency virus infection in North American women. *Medicine* 70 (5):307-326.
Filice, G.A. and C. Pomeroy. 1991. Preventing secondary infections among HIV-positive persons. *Public Health Reports* 106, no.5 (September-October):503-18.
Fritzler, M.J. and M. Salazar. 1991. Diversity and origin of rhematologic autoantibodies. *Clinical Microbiology Reviews* 4 (3):256-69.
Landay, A.L., C. Jessop, E.T. Lennette, and J.A. Levy. 1991. Chronic fatigue syndrome: clinical condition associated with immune activation. *Lancet* 338, no.8769 (September 21):707-13.
Mann, J. 1991. AIDS and the next pandemic. *Scientific American* 264 (March):126.
Mills, J. and H. Masur. 1990. AIDS-related infections. *Scientific American* 263 (August):50-57.
Steere, A.C., E. Dwyer, and R. Winchester. 1990. Association of chronic lyme arthritis with HLA-DR4 and HLA-DR2 alleles. *New England Journal of Medicine* 323, no.4 (July 26):219-23.
Steffy, K. and F. Wong-Staal. 1991. Genetic regulation of human immunodeficiency virus. *Microbiological Reviews* 55 (2): 193-205.
Stiff, J., M. McCormack, E. Zook, T. Stein, and R. Henry. 1990. Learning about AIDS and HIV transmission in college-age students. *Communication Research* 17, no.6 (December):743-59.

Chapter 20: Diseases of the Skin and Eyes; Wounds and Bites

Anonymous. 1991. Increase in rubella and congenital rubella. *The Journal of the American Medical Association* 265, no.9 (March 6):1076-77.
Anonymous. 1991. Treatment of ocular herpes zoster. *Lancet* 338, no.8777 (November 16):1244-46.
Dart, J.K.G., F. Stapelton, and D. Minassian. 1991. Contact lenses and other risk factors in microbial keratitis. *Lancet* 338, no.8768 (September 14):650-54.
Manetti, A.C. 1991. Hyperendemic urban blastomycosis. *American Journal of Public Health* 81, no.5 (May):633-37.
Marwick, C. 1991. Measles eradication? Data suggest reaching goal will be a challenge. *The Journal of the American Medical Association* 265, no.17 (May 1):2163.
Taylor, H.R., J.A. Siler, H.A. Mkocha, B. Munoz, V. Velez, L. DeJong, and S. West. 1991. Longitudinal study of the microbiology of endemic trachoma. *Journal of Clinical Microbiology* 29 (8):1593-95.

Weingardt, J. and Y. Li. 1991. North American blastomycosis. *American Family Physician* 43, no.4 (April):1245-49.

White, G.L. Jr., S.M. Thiese, H.E. Olafsoon, and M.K. Lundergan. 1991. Disposable contact lenses. *American Family Physician* 43, no.5 (May):1643-47.

Zoutman, D.E. and L. Sigler. 1991. Mycetoma of the foot caused by Cylindrocarpon destructans. *Journal of Clinical Microbiology* 29 (9):1855-59.

Chapter 21: Diseases of the Respiratory System

Anonymous. 1991. A pictorial journey through humanity's battle against the white plague. *Nutrition Health Review* 59 (Summer):8-9.

Anonymous. 1991. Robert Koch discovers the TB bacillus. *Nutrition Health Review* 59 (Summer):6.

Anonymous. 1991. TB and deer farming: return of the king's evil? *Lancet* 338, no.8777 (November 16):1243-45.

Bell, T.A., W.E. Stamm, S.P. Wang, C.C. Kuo, D.D. Holmes, and J.T. Grayston. 1992. Chronic Chlamydia trachomatis infections in infants. *The Journal of the American Medical Association* 267, no.3 (January 15):400-403.

Ellis, D.H. and T.J. Pfeiffer. 1990. Natural habitat of Cryptococcus neoformans var. gattii. *Journal of Clinical Microbiology* 28 (7):1642-44.

Fanning, A. and S. Edwards. 1991. Mycobacterium bovis infection in human beings in contact with elk (Cervus elaphus) in Alberta, Canada. *Lancet* 338, no.8777 (November 16):1253-56.

Halstead, D.C., S. Todd, and G. Fricth. 1990. Evaluation of five methods for respiratory syncytial virus detection. *Journal of Clinical Microbiology* 28 (5):1021-25.

Kellogg, J.A. 1990. Suitability of throat culture procedures for detection of Group A streptococci and as reference standards for evaluation of streptococci antigen detection kits. *Journal of Clinical Microbiology* 28 (2):165-69.

Lefler, E., D. Weiler-Ravell, D. Merzbach, O. ben-Izhak, and L.A. Best. 1992. Traveller's coccidiodomycosis: case report of pulmonary infection diagnosed in Israel. *Journal of Clinical Microbiology* 30 (5):1304-06.

Lundren, B., J.A. Kovacs, N.N. Nelson, F. Stock, A. Martinez, and V.J. Gill. 1992. Pneumocytis carinii and specific fungi have a common epitope, identified by a monoclonal antibody. *Journal of Clinical Microbiology* 30 (2):391-95.

Thomas, P. 1991. Common Misery. *Harvard Health Letter* 17, no.1 (November):6-9.

Yakrus, M.A. and R.C. Good. 1990. Geographic distribution, frequency, ans specimen source of Mycobacterium avium complex serotypes isolated from patients with acquired immuno-deficiency syndrome. *Journal of Clinical Microbiology* 28 (5):926-29.

Chapter 22: Oral and Gastrointestinal Diseases

Anonymous. 1991. Spirochaetes in periodontal disease. *Lancet* 338, no.8776 (November 9):1177-78.

Anonymous. 1991. Update: Cholera - Western Hemisphere, and recommendations for treatment of cholera. *The Journal of the American Medical Association* 266, no.9 (September 4):1186-87.

Clayton, C.L., H. Kleanthous, P.J. Coates, D.D. Morgan, and S. Tabaqchali. 1992. Sensitive detection of Helicobacter pylori by using polymerase chain reaction. *Journal of Clinical Microbiology* 30 (1):192-200.

Current, W.L. and L.S. Garcia. 1991. Cryptosporidiosis. *Clinical Microbiology Reviews* 4 (3):325-58.

Glass, R.I., M. Claeson, P.A. Blake, R.J. Waldman, and N.F. Pierce. 1991. Cholera in Africa: lessons on transmission and control for Latin America. *Lancet* 338, no.8770 (September 28):791-95.

Jones, D. 1990. Foodborne listeriosis. *Lancet* 336, no.8724 (November 10):1171-75.

Lewis, D.L. and R.K. Boe. 1992. Cross-infection risks associated with current procedures for using high-speed dental handpieces. *Journal of Clinical Microbiology* 30 (2):401-06.

Skirrow. 1990. Campylobacter (foodborne illness). *Lancet* 336, no.8720 (October 13):921-23.

Soltesz, V., B. Zeeberg, and T. Wadstrom. 1992. Optimal survival of Helicobacter pylori under various transport conditions. *Journal of Clinical Microbiology* 30 (6):1453-56.

Tucker, K.D., P.E. Carrig, and T.D. Wilkins. 1990. Toxin A of Clostridium difficile is a potent cytotoxin. *Journal of Clinical Microbiology* 28 (5):869-71.

Valentine, J.L., R.R. Arthur, H.L.T. Mobley, and J.D. Dick. 1991. Detection of Helicobacter pylori by using the polymerase chain reaction. *Journal of Clinical Microbiology* 29 (4):689-95.

Wolford, K. 1991. Peru in the time of cholera. *Christian Century* 108, no.30 (October 23):969-71.

Chapter 23: Cardiovascular, Lymphatic, and Systemic Diseases

Anderson, J.F., E.D. Mintz, J.J. Gadbaw, and L.A. Magnarelli. 1991. Babesia microti, human babesiosis, and Borrelia burgdorferi in Connecticut. *Journal of Clinical Microbiology* 29 (12):2779-83.

Anonymous. 1991. Trichinosis outbreaks. *FDA Consumer* 25, no.4 (May):4-5.

Anonymous. 1991. WHO vows to eradicate leprosy by year 2000. *UN Chronicle* 28, no.3 (September):64.

Blacklow, N.R. and H.B. Greenberg. 1991. Viral gastroenteritis. *New England Journal of Medicine* 325, no.4 (July 25):252-65.

Dennis, D.T. 1991. Lyme disease: tracking an epidemic. *The Journal of the American Medical Association* 266, no.9 (September): 1269-70.

Dolan, P.J., R.M. Skibba, R.C. Hagan, and W.R. Kilgore. 1991. Hepatitis C: prevention and treatment. *American Family Physician* 43, no.4 (April):1347-55.

Gasser, R.A., A.J. Magill, C.N.Oster, and E.C. Tramont. 1991. The threat of infectious disease in Americans returning from Operation Desert Storm. *New England Journal of Medicine* 324, no.12 (March 21):859-65.

Guerrant, R.L. and D.A. Bobak. 1991. Bacterial and protozoal gastroenteritis. *New England Journal of Medicine* 325, no.5 (August 1):327-40.

Hoofnagle, J.H. 1990. Chronic hepatitis B. *New England Journal of Medicine* 323, no.5 (August 2):337-40.

Okano, M., S. Matsumoto, T. Osato, Y. Sakimaya, G.M. Thiele, and D.T. Purtilo. 1991. Severe chronic active Epstein-Barr virus infection syndrome. *Clinical Microbiology Reviews* 4 (1): 129-35.

Szer, I.S., E. Taylor, and A.C. Steere. 1991. The long-term course of Lyme arthritis in children. *New England Journal of Medicine* 325, no.3 (July):159-63.

Tiollais, P. and M. Buendia. 1991. Hepatitis B virus. *Scientific American* 264 (April):116-20.

Wadsworth, L.T. 1991. Rheumatic fever is back: don't miss it. *Physician and Sports Medicine* 19, no.4 (April):75-81.

Chapter 24: Diseases of the Nervous System

Anonymous. 1990. Botulinum toxin approved for treating eye disorders. *ASM News* 56 (3):132-33.

Cartwright, K.A.V., D.M. Jones, A.J. Smith, J.M. Stuart, E.B. Kacznmarski, and S.R. Palmer. 1991. Influenza A and meningococcal disease. *Lancet* 338 (8766):554-59.

Fischetti, M. 1991. Gas vaccine: bioengineered immunization could shield against nerve gas. *Scientific American* 264 (April): 153-54.

Gellin, G.B. and C.B. Broome. 1989. Listeriosis. *The Journal of the American Medical Association* 261, no.9 (March 3):1313-21.

Guinn, B. 1983. Hansen's disease in south Texas. *Health Education* 14, no.1 (January-February):46-51.

Heneson, N. 1991. New clinical uses for botulinum toxin. *ASM News* 57 (2):63-64.

Holloway, M. 1991. Neural vector: herpes may open the way to gene therapy in neurons. *Scientific American* 264 (January):32.

Roche, P.W., W.J. Theuvenet, and W.J. Britton. 1991. Risk factors for type-1 reactions in borderline leprosy patients. *Lancet* 338, no.8768 (September 14):654-58.

Theccanat, G., L. Hirschfield, and H. Isenberg. 1991. <u>Serratia marcescens</u> meningitis. *Journal of Clinical Microbiology* 29 (4):822-23.

Chapter 25: Urogenital and Sexually Transmitted Diseases

Anonymous. 1991. Combination antimicrobial therapy in the treatment of acute pelvic inflamatory disease. *The Journal of the American Medical Association* 266, no.18 (November 13):2542.

Anonymous. 1991. Genital human papillomavirus infection. *American Family Physician* 43, no.4 (April):1279-91.

Cowley, G. 1991. Sleeping with the enemy. *Newsweek* 118, no.24 (December 9):58-60.

Dan, B.B. 1990. Sex, lives, and chlamydia rates. *The Journal of the American Medical Association* 263, no.23 (June 20):3191-92.

Handsfield, H.H. 1990. Old enemies: combating syphilis and gonorrhea in the 1990's. *The Journal of the American Medical Association* 264, no.11 (September 19):1451-53.

Johnson, J.R. 1991. Virulence factors in <u>Escherichia coli</u> urinary tract infection. *Clinical Microbiology Reviews* 4 (1):80-128.

Kaufman, R.H. 1991. Human papillomavirus and cervical cancer: risk to male partner. *The Journal of the American Medical Association* 265, no.9 (March 6):1179-80.

Merkus, J.M.W.M. 1990. Treatment of vaginal candidiasis: orally or vaginally? *Journal of the Americn Academy of Dermatology* 23, no.3 (September):568-73.

Morgan, R.J. 1991. Clinical aspects of pelvic inflammatory disease. *American Family Physician* 43, no.5 (May):1725-33.

Smith, S.M., T. Ogbara, and R.H.K. Eng. 1992. Involvement of <u>Gardnerella vaginalis</u> in urinary tract infections in men. *Journal of Clinical Microbiology* 30 (6):1575-77.

Sherman, K.J., J.R. Daling, J. Chu, N.S. Weiss, R.L. Ashley, and L. Cory. 1991. Genital warts, other sexually transmitted diseases, and vulvar cancer. *Epidemiology* 2 (4):257-62.

Spiegel, C.A. 1991. Bacterial vaginosis. *Clinical Microbiology Reviews* 4 (4):485-502.

Washington, A.E., S.O. Aral, P. Wolner-Hanssen, D.A. Grimes, and K.K. Holmes. 1991. Assessing risk for pelvic inflammatory disease and its sequelae. *The Journal of the American Medical Association* 266, no.18 (November 13):2581-87.

Chapter 26: Environmental Microbiology

Anonymous. 1990. An ocean of viruses may affect global cycles. *ASM News* 56 (12):632-33.

Aldhous, P. and K.S. Jayaraman. 1991. Big test for bioremediation. *Nature* 349, no.6309 (February 7):447.

Alper, J. 1990. Oases in the oceanic desert. *ASM News* 56 (10): 536-38.

Gillis, A.M. 1990. Why deserts lose nitrogen. *BioScience* 40 (November):716-18.

Hantula, J., A. Kurki, and P. Vuoriranta. 1991. Ecology of bacteriophages infecting activated sludge bacteria. *Applied and Environmental Microbiology* 57 (August):2147-51.

Holloway, M. 1991. Soiled shores. *Scientific American* 265 (October):102-106.

Junior, O.G. 1991. Isolation and purification of <u>Thiobacillus ferroxidans</u> and <u>Thiobacillus thiooxidans</u> from some coal and uranium mines of Brazil. *Rev. Microbiol* 22 (1):1-6.

Lewis, K.L. and D.K. Gattie. 1991. The ecology of quiescent microbes. *ASM News* 57 (1):27-32.

Madsen, E. 1991. Determining in situ biodegradation; facts and challenges. *Enivronmental Science & Technology* 25 (October):1662-73.

Odom, J.M. 1990. Industrial and environmental concerns with sulfate-reducing bacteria. *ASM News* 56 (9):473-76.

Richmond, J.Y. 1990. Developing a response to medical waste issues. *ASM News* 56 (8):420-23.

Song, H., X. Wang, and R. Bartha. 1990. Bioremediation potential of terrestrial fuel spills. *Appl. Environ. Microbiol* 56 (3): 652-56.

Chapter 27: Applied Microbiology

Anonymous. 1990. Brewing a sausage. *Lancet* 336, no.8730 (December 22):1546-48.

Beardsley, T. 1991. A nitrogen fix for wheat. *Scientific American* 264 (March):32.

Coghlan, A. 1991. An explosive start to fast-maturing cheeses. *New Scientist* 129, no.1760 (March 16):28.

Fleet, G. 1992. Spoilage yeasts. *Crc Crti. Rev. Biotechnol.* 12 (12):1-44.

Fliss, I., R.E. Simard, and A. Ettrki. 1991. Comparison of three sampling techniques for microbiological analysis of meat surfaces. *Journal of Food Science* 56 (January-February): 249-50.

Franklin, M.J. and D.C. White. 1991. Biocorrosion. *Current Opinion in Biotechnology* 2 (3):450-56.

Maigetter, R.Z., F.J. Bailey, and B.M. Miller. 1990. Biosafety in a large-scale fermentation laboratory. *ASM News* 56 (2):82-86.

Mise, K. 1990. Application of recombinant DNA technology to food microbiology. *J. Food Hyg. Soc. Japan* 31 (3):209.

Salminen, S., S. Gorbach, and K. Salminen. 1991. Fermented whey drink and yogurt-type product manufactured using <u>Lactobacillus</u> strain. *Food Technology* 45, no.6 (June):112.

Watts, S. 1990. Have we the stomach for engineered food? *New Scientist* 128, no.1741 (November 3):24-26.

Williams, R.P. and A.L. Gillen. 1991. Microbe phobia and kitchen microbiology. *The American Biology Teacher* 53 (January):10-11.